Instructor's Resource Manual

Connie Re Rothwell
*University of North Carolina
at Charlotte*

THE BLAIR READER

Third Edition

LAURIE G. KIRSZNER
*Philadelphia College
of Pharmacy and Science*

STEPHEN R. MANDELL
Drexel University

PRENTICE HALL, Upper Saddle River, New Jersey 07458

CONTENTS

iv

INTRODUCTION

Teaching with

The Blair Reader, third edition

Using This Instructor's Resource Manual

This *Instructor's Resource Manual* is designed to help you and your students get the most out of *The Blair Reader*, third edition. In order to take full advantage of this resource, you might want to familiarize yourself with the different components of this manual. As you will see, the advantage of teaching with *The Blair Reader* is that you can use it to meet various teaching goals. It is our hope that this manual will give new ideas to all teachers whether you are just beginning your teaching career or looking for ways to renew your practice. We hope that this manual will also save you in planning time so that you have more time for your students' individual needs. Most of all, we hope that the ideas suggested in this manual will help you and your students engage in reading and writing, and help you enjoy your course.

Organizational Principles of This Manual

Each of the book's ten chapters follows the same organizational principles. We recommend that you familiarize yourself with this organization before creating your syllabus.

Chapter Prompts

Setting Up the Unit: Each chapter of this manual begins with a discussion of a possible strategy teachers can use in their treatment of that chapter. Each strategy described in these sections is based on techniques, approaches, structures, or appeals suggested by a number of the selections in that chapter. Strategies for each chapter are as follow:

Confronting the Issues: Each chapter also begins with a range of activities designed to unearth and explore some of the issues suggested by the readings that follow. Most are collaborative activities, useful for facilitating student-centered classes; many translate easily into fuller-length projects, for those students who are intrigued by the possibilities opened up by the activity. There are four types of *Confronting the Issues* prompts:

Constructing Contexts: These are generally discussion-based suggestions, often with supplementary information the teacher provides. Most work to expose students to a variety of pressing issues related to a single thematic area; many ask students to reach some consensus through discussion and group work.

Community Involvement: These projects require students to learn more about their local and college communities, expanding the walls of the classroom to include the community at large. Each

vii

project includes a collaboratively written document that students can use as a starting point for a longer research and writing project.

Cultural Critique: These prompts encourage students to investigate the cultural and social pressures that help them construct their identities. Often their analysis of images leads to an explication of the ways in which we all interact with and intervene in those processes of identity formation.

Feature Films: These prompts suggest full-length movies as springboards for discussions in which students analyze the visual text and relate their critique to the selections in the chapters and to their lives. Most of the films are widely available at public libraries or local video stores; some may be owned by your school's library or audiovisual department.

Teaching "Two Perspectives" : Each chapter of the textbook begins with two readings that have been paired because they create a context for an issue. We have provided a pre-reading exercise for each of these paired readings, asking students to consider their own views on a topic before reading the views of others. *Two Perspectives: Suggested Answers for Responding to Reading Questions* appear at the end of the individual annotations for the selections.

Focus: The set of essays at the end of each chapter narrows the chapter's theme to a particular issue or question. This section encourages critical reading and writing by asking students to respond to different claims on the same issue or question. The prompts in this section help students to integrate their readings to form their own opinion.

Selection Prompts

For Openers : The first prompt for each selection is designed to open class discussion after the students have read the piece. Most are best suited for a full-class discussion, but you can have students begin discussion in small groups. Alternatively, you can ask students to freewrite on the question before discussion begins, in order to give them a chance to formulate their responses or reflect on their preconceptions.

Teaching Strategies: These suggestions are designed to stimulate students' discussions, facilitate their understanding, guide their reading, or provide additional information on the writer or the reading selection. For example, we may suggest close reading exercises or suggest ways to question cultural assumptions that the writer seems to be making.

Collaborative Activities: Through these activities, students can work together to make meaning out of the readings and develop insight and understanding by exchanging ideas with others—both inside and outside the classroom. These prompts suggest possibilities for small-group interaction and collaboration, sometimes resulting in individual group projects, sometimes leading to full-class discussions.

Writer's Options: Each reading selection offers possibilities for exploring the reader's personal responses, the writer's stylistic devices, and the text's thematic issues—either in a journal entry or in a longer, more structured writing assignment. These prompts offer suggestions for such writing projects.

Multimedia Resources: By looking at the other "texts" besides those contained between the covers of the textbook, students can expand their understanding of the issues raised by a particular selection. For almost every reading, we offer suggestions for introducing students to movies, songs, television programs, cartoons, merchandise, advertisements, and Internet sources. These supplementary texts either address the main selection by filling in additional background or they present a challenge to the main text through a critical analysis of the issue at hand.

Suggested Answers for Responding to Reading Questions: Each individual reading selection is followed by three questions for students. We provide some possible responses in this Manual.

Additional Questions for Responding to Reading: For each selection, we have also included at least two, and sometimes as many as five, additional questions for those instructors who would like a little more editorial apparatus than the textbook contains. In many cases, these questions could be used as journal or paper topics or as exam questions.

Guidelines for Building a Syllabus

The Blair Reader, third edition, permits you to follow the textbook as closely — or as tangentially — as you like. Discussed below are some options for teaching with *The Blair Reader*. Browse through them, and see which of the strategies suit your pedagogical approach.

Perhaps the most obvious way to teach from any reader is to use the selections within the book as sources for class discussions and papers. Whether you use readings to introduce themes, to illustrate rhetorical models, to present differing perspectives, or even to test comprehension, you will find *Questions* within the textbook that address these concerns. Since students also have access to these *Questions*, you can easily assign them along with the reading to help facilitate thoughtful classroom discussions. In addition to the prompts with the student editions, this *Instructor's Resource Manual* includes at least two *Additional Questions* that you can use to guide students in their critical reading. You can supplement your discussions or homework assignments with these questions.

There are a number of different ways to use *The Blair Reader* in a thematically structured course. You can begin by selecting the chapters whose topics seem most appropriate for your students and assign readings that comment on one another. One place to start is with the *Two Perspectives* pairs that open each chapter and the *Focus* essays that end each chapter. Another possibility is to look through the *Topical Clusters* at the beginning of the book; some of those narrow themes might work better in your class than the broader chapter themes. To provide additional contextual perspectives, we have listed *Multimedia Resources* for most selections. These songs, videos, movie scenes, Internet sources or supplementary print readings comment on the primary text in some way. If you are interested in teaching rhetorical principles, you can consult the *Rhetorical Table of Contents* that follows the book's main *Table of Contents* (in which readings are arranged thematically).

Teaching Student Writing

Each chapter in the book begins with *Student Voices*, excerpts from journal entries written by students using *The Blair Reader*. You can use these informal student responses as springboards for opening discussion, as models for replication, or as points for dissention. This section serves as an example of an important part of the reading and writing process that should not be taken for granted: the student's own experiences, biases, and culture that he or she brings to any study. The shape of an idea begins to take shape with one's own initial jottings which change and take shape with continued reading, discussion, and writing. Do spend some time drawing your students into conversation with these other students. Be sure students understand that one does not enter academic inquiry in silence.

Teaching with Journals

For every chapter, and every reading within chapters, the *Instructor's Resource Manual* offers suggestions for journal entries. Some are self-contained; others can easily be expanded into longer projects. All have been designed to help students move from reading to writing — by helping them to draw connections between their experiences and the subjects of the essays, asking for descriptions, memories, or reactions. A journal can provide students with a "safe" place to try writing, or it can build a stronger understanding (through written dialogues) between student and instructor (or even between students in journal groups). Journals can also be easily done on-line. Finally, students can use pieces of their journal entries in their papers. For more specific advice on teaching with journals, see *Guidelines for Working with Journals* below.

Teaching within Communities

If you choose to have your students write with a strong connection to their communities (either the college community or the larger geographical area), supplement the selections in *The Blair Reader* with your school or town newspaper. In addition, each chapter of the manual includes a *Confronting the Issues* option called *Community Involvement*. These require that students go into their communities to learn, to teach, to suggest possible solutions to problems they identify, or to find out more about the resources that a community does (or does not) offer. Most of these explorations are designed to result in a collaboratively written document addressing particular needs of a community. Students may gain a real appreciation for ways academic inquiry can have a real world effect on the school or area community.

Teaching against the Grain

For those of you interested in engaging your students in cultural critique throughout the semester, look particularly at the selection prompts for *Multimedia Resources*. These annotations often ask students to link a selection's key terms and assumptions with related "texts," suggesting connections to issues in students' (and teachers') everyday lives. Many of the prompts in *Teaching Strategies* also ask students to identify the values and assumptions of the writer or narrator, exposing gaps (or parallels) between reader and writer in the process. Finally, two options within the section suggest possibilities for reading into culture: *Cultural Critique* asks students to analyze images and myths pertaining to the topic of the *Confronting the Issues* chapter, and *Feature Films* asks students to view a full-length presentation touching on relevant issues. For additional advice, see *Guidelines for Using Multimedia Resources in the Classroom* below.

Teaching without the Reader

This may seem like a strange direction from somebody who has already adopted *The Blair Reader* as a textbook, but we realize that quite often the decision maker is not the instructor using the book. If you prefer not to use this text to structure your syllabus, you can still use *The Blair Reader* as a departure point for determining the shape your class will take. Start with a selection or a prompt within this guide, and see where it takes you and your students. Use some of the *Confronting the Issues* options to create class projects. Or, start the course in a completely different direction, having students dip into the textbook when and where they are interested and suggest selections they would like to bring to a class discussion; then, build your syllabus around those choices. The goal of both the reader and this Manual is to provide you with as many resources and ideas as possible, but it is up to you to shape them into a pedagogy that is responsive to your students and responsible to yourself. And, after all, you know your teaching style, your students, and your institution better than we ever could.

Teaching Anywhere, Any Way

No matter how you teach, strive to get to know your students and work with them to create the syllabus for your class. For possible ways to learn more about your students quickly at the beginning of the semester, see the section *Guidelines for Conducting Needs Assessment* below. You may want to plan only one unit at a time, as you gradually get to know the values, expectations, desires, and needs of your individual class or classes. With *The Blair Reader* and this *Instructor's Resource Manual*, you can be as flexible as you need to be throughout the semester, and you can be as responsive as you would like to be to each class of students you teach.

Guidelines for Conducting Needs Assessment

One of the best ways to "meet students where they are" is to make an effort in the first week of class to find out "where" they are. Often, time constraints do not permit much opportunity for you to assess your students' strengths, abilities, and interests along with their needs and deficiencies. Still, it is important to try.

Many programs require teachers to administer a writing sample within the first week of class, partly to see if students have been placed in an appropriate class level, and partly to have a hand written, supervised writing sample on file to guard against future plagiarism. This writing sample can also provide individual instructors with a wealth of information on each student. For example, you can establish records of the strengths and weaknesses of individual writers, or you can identify common needs in your students which might then be addressed in a class session. Likewise, if there are issues that you usually teach, but that this particular group of students already has under control, you needn't take class time to discuss them.

Also, if you have control over the topic of the writing sample, you can use it to begin exploring themes that might run through your class, or the topic of the first paper. In other words, find out student ideas on a subject at the beginning of the term to see what values and opinions are represented within your class; this allows you to choose selections from the textbook to complement their interests and perspectives. Thus, these writing samples can serve as a base line for assessing your students' inclinations as well as their abilities.

Gathering biographical information that can illuminate the backgrounds and perceptions that students bring with them to class is also very useful. In addition to collecting phone numbers and advisors' names, ask students about their previous writing instruction. What did they find most helpful? Least helpful? Have they received instruction in organization, style, pre-writing, persuasive strategies? What do they perceive as their greatest strength as a writer? Their greatest weakness? What would they like to improve in the course of the semester? What kinds of writing have they done besides academic writing?

A more formal way to assess students' attitudes toward their own writing is to administer the twenty-six question *Daly-Miller Writing Apprehension Inventory*. This resource measures students' level of anxiety about writing, and it can be done using a scannable form. Reliability is quite high, and the test can tell instructors how eager or reluctant their students might be. It will also identify students who might need some special attention in order to get over severe writer's block.

If time allows, you can find out much of this information through individual conferences with students. Unfortunately, students won't get to know each other very well in this process. By using in-class techniques, you can ask students to indicate responses to certain questions, sharing parts of their backgrounds with classmates. Keep in mind, though, that although it might be tempting to play "get-to-know-you" games with students on the first day, many teachers find that these backfire; many students feel they are patronizing, and they can send a signal that a composition class is not intellectually rigorous. We recommend other ways of learning about your students.

Throughout the semester, you can gauge the progress your students feel they are making by requesting that they submit a "Writer's Memo" with their first drafts. These memos can guide the way you respond to your students' writing. Have them tell you what kind of advice they received from their peer groups and what they hope to revise, what sorts of help they want from you, what they perceive to be the best part of their paper and the weakest.

Guidelines for Working with Collaborative Groups

Many instructors work with some form of peer interaction, whether as peer discussion, peer review of papers, or co-authored papers. Groups of three to five students are most productive, but group size depends on the size of your class, the scope of your project, and the type of peer activity you want to assign. Co-authored papers (or sections of a larger all-class project) tend to work best in smaller groups, and time constraints limit the numbers of peer reviews that can be covered in a fifty-minute class. In these situations, you probably won't want more than three students to a group. On the other hand, for in-class discussions, ethnographic research, and brainstorming sessions, larger groups — say, five students — help expose a wider variety of opinions and backgrounds, making premature consensus more difficult to achieve.

Opinions differ on the ideal student composition of collaborative groups. Some teachers feel that homogenous groups work better, with students of roughly equal capabilities pushing each other to achieve more. Group members whose writing quality differs too greatly, the argument goes, pull the group apart, intimidating the less able students and lowering the standards of those with better previous instruction. On the other hand, there is a great deal to be said for getting to know students from disparate backgrounds and levels of achievement, and group work is a good way to explore issues of difference through collaboration and cooperation. (And early in the term, of course, you will not know enough about your students to create precisely homogenous groups.)

A similar debate concerns the gender distribution of collaborative groups. Some teachers find that female students, in particular, can be reluctant to participate in mixed-gender groups, whereas they sometimes flourish in all-female groups. This problem can be diminished by placing more females than males in a group, but this is rarely possible for an entire class. And, again, the chances for having a varied dialogue is greatly diminished if the sexes are segregated.

Regardless of the ways in which you decide to assign students to groups (or allow them to self-select), in-class group time is best spent when the instructor clearly defines goals, provides guidelines, and sets a time limit for collaborative efforts. Write the list of tasks they are to accomplish together on the board to help keep them focused. Without over-determining the end result for your students, suggest how much time they should try to spend on each task in a project, writing the times on the board for their reference. By far the largest complaint about group work from teachers is that it often degenerates into social hours; the suggestions above might help alleviate that problem and help students get more out of their time together. However, don't be too distressed by some chatting. Students need to understand each others' ways of communicating and must develop trust for group work and for peer revision Some informal conversation can help build trust.

If students in groups are asked to respond to drafts of their classmates' papers, it is helpful to provide them with specific guidelines, a set of perhaps five questions to answer about the paper they are reading. (For example: Is the paper clearly organized? Are transitions provided? Does the paper state and support a main idea? Does it include enough detail? Is it appropriate for its audience? Does it fulfill the assignment?) Students can do more than respond to these questions, of course, but they should not do less. The number-one complaint about peer responses by teachers is that students do not address issues of content or organization, focusing instead on grammatical and stylistic issues that are best dealt with in later draft stages. It is important, then, to model good responses yourself, and to spend at least one class period helping students learn to respond to each other in ways that will be useful — perhaps discussing two or three drafts as a class.

Finally, when relying on groups to co-author papers, it is very important to allow students a chance to evaluate themselves and their peers. Often students will feel that certain members didn't pull their share of the duty, or that the process of collaboration was more frustrating than productive. Both are potential pitfalls of collaboratively written papers, and students will need to articulate for themselves what they think of the process — what they learned, what they would replicate, and what they would do differently the next time. These responses can help you evaluate individual work in a group project and they will alert you to potential problems the next group of students might encounter.

Guidelines for Working with Journals

Using writing and reading journals as an aid to comprehension, invention, and drafting is a common pedagogical tool. Thousands of teachers use journal writing in one form or another in their composition classes, with various goals and requirements. Some feel that the more students write, the better; journals in these classes are often completely open-ended and graded on the basis of page length alone. Other instructors use them as a way to have students begin to articulate their responses to issues that come up in class. Some teachers have students practice stylistic techniques; others ask them to freewrite for fifteen minutes a day without stylistic or grammatical concerns. The possibilities are too numerous to describe here.

You will, however, need to make some decisions before requiring journal assignments for your students. The main issue is the level of privacy that you want to allow students. Some teachers prefer to have teams of students read one another's journals, commenting and creating a conversation among students rather than between student and teacher. Other instructors like to check the journals regularly, even after every assignment, in order to provide constant feedback, encouragement, and additional ideas; the journal, then, becomes an extended dialogue between student and teacher. Others request that students submit a minimum number of pages to be read, blocking out or covering up anything that they don't want to share. Still others never look at journals at all. The choice is yours, one you may wish to make in consultation with your students; regardless of your decision, make sure students understand your policies and your expectations.

Guidelines for Evaluating
Student Writing

The primary rule of thumb in evaluating student writing is to be clear about your criteria for evaluation—and consistent in carrying out those criteria. Make sure students understand how their grade is determined. For example, do you place any emphasis on invention strategies before the first draft? On participation in class discussion or group projects? On quantity and quality of revisions? Many students may assume their grade is calculated on the basis of the final product alone, so if your approach encourages and rewards the full process, make sure the students understand this from the outset.

Some instructors include their students in the creation of the evaluation criteria. They begin by asking students what they think constitutes a good paper, and students help assign relative merit to the various parts of a paper. Such discussions provide an opportunity for you to teach students about your own view of writing and what kinds of learning can be accomplished through composition instruction. Some of your ideas might be new to many students, so be prepared to explain clearly and thoughtfully. Students may be more willing to achieve standards they helped set than those they see as externally (and perhaps arbitrarily) imposed. It is also helpful for students if you attach an explanation of how they did—or did not—fulfill the criteria that had been agreed upon. Not only will students be less likely to appeal for a higher grade, they will have a better understanding of the ways in which they can continue to improve.

Portfolios are becoming an increasingly popular way to evaluate student writing. Rather than assigning point values to papers as students turn them in throughout the semester, you may evaluate student portfolios at the end of the semester in a coherent unit. This approach gives students time to reflect upon work they did earlier in the semester, revising it (as time permits) in light of insights they gained after the paper would have been turned in. Portfolios reinforce the idea of writing as an ongoing process and the belief that students can learn through writing.

Although time constraints and large class size often preclude numerous individual conferences with students throughout the semester, the time is well spent. It is very difficult to know how students will react to written comments on their papers, so using conferences as an integral part of evaluation can give students a chance to ask questions—and can give you a chance to clear up any misunderstandings. Even the most conscientious instructors may write comments on papers that can be taken in a way that is very different from the one intended, and increased personal contact can lessen the potential for confusion or hurt feelings.

Guidelines for Using Multimedia
Resources in the Classroom

In response to the increasing number of teachers who want to incorporate some form of cultural critique into their classes, we have included over one hundred possibilities for teaching with alternative forms of media. We have suggested ways to work with certain films, songs, cartoons, commercial products, and so on—and, of course, you may have ideas of your own. Regardless of how you use media in the classroom, you might want to keep a few guidelines in mind.

Be sure to plan ahead. Bringing in sources from outside the classroom can take more time than you think, so fit that time into your schedule. If you are renting a movie to show to your class, order it at least two days earlier than you need it; in case it has been checked out by somebody else, you will still have time to reserve it when it is returned, or to find other sources.

Likewise, if you need special equipment that your classroom does not already contain, you will need to contact the appropriate campus department and request it. If they do not have the equipment, you will need time to make other arrangements. For example, many video rental stores now also rent VCRs for a nominal fee. If you need a tape player, you might need to ask a student to bring in a portable stereo.

Before the semester begins, it is a good idea to call various departments within your school to see what options you might have available. Some institutions have sophisticated equipment housed in departments other than your own (laser disc players, computerized transparency machines, CD-ROM machines, etc.) that you might be able to borrow. Other departments may allow you to use their resources, including multimedia classrooms, on a limited basis. Psychology and music departments are often well-stocked. If you do some research before the start of the term, you can save yourself a lot of time and trouble.

You will save yourself money as well as time by researching the available materials at your school and local libraries. Many loan videos, CDs, magazines, and any number of other resources. Make an appointment to meet with a reference librarian to find out the scope of the possibilities. Clearly, renting several movies can eat into your personal budget, and not many instructors are allotted departmental resources for class materials.

Because of the effort involved in setting up movie screenings and splicing together illustrative video or audio tapes, you might want to include other teachers in your planning. Several of us have collaborated on shared movie viewings, requesting a large, comfortable amphitheater with a big screen. Likewise, during a presidential election year, we took turns taping and creating compilations of the public debates. Chances are, there are other instructors in your department or school who would like to collaborate with you.

Finally, involve your students in the process. They may own a wealth of resources at their homes or in their dorms. Encourage them to bring additional materials to class; you can even assign them to create compilations of tapes that you can use in future semesters. The possibilities are endless.

CHAPTER 1

FAMILY AND MEMORY

Setting Up the Unit: Using Personal Writing to Enrich Exposition

In this first chapter of *The Blair Reader*, third edition, students can read about writers from diverse backgrounds who have interpreted their experiences in ways quite similar to those of your students. Readers should see that some of the writers' attitudes and reactions are comparable to their own despite the wide variation in the demographics of the authors in this chapter. At the same time however, they should understand that differing backgrounds and points of view do alter the ways in which the same experience might be lived, remembered, and described by different people or at different stages within the same person's life.

As we state in the introduction to this chapter, the writings that follow are attempts to understand, to recapture, to recreate pieces of memories that can contribute to the writer's sense of self. Through their memories of family interactions, these authors write about their attempts to construct a coherent identity for themselves out of the fragments of their own understanding and their perceptions of others. The different writers approach the subject of identity formation in a number of ways, but with at least one common element: All stress the ways in which our self-identity is constructed socially, through the interactions of home, work, school, extended family, friends, religion, socioeconomic class, gender, race, and so on. These writers do not identify themselves as fitting into one category and one category only. Rather, many selections here trace a writer's developing awareness of his or her identity through a variety of family and social interactions. In all of the selections, *home* and *family* are important components in the shaping and discovering of a sense of self; however, almost every piece defines these concepts of home and family in very personal and distinct ways.

Because of this emphasis on family and self-awareness, many of the readings in this chapter are autobiographical, leading students to write their own interpretations of their own experiences. Many instructors like to begin the semester with personal writing, assuming students can start the semester comfortably with subjects they know well—themselves, their families, and their homes. These subjects are accessible, yet rich with possibility. Students thus begin the course from a position of strength.

Because students may not know themselves as well as they like to think they do, autobiographical writing can also provide an opportunity to gain new insight into subjects that may seem commonplace. Personal writing should encourage students to learn something new about themselves through the process of thinking and writing. Their knowledge may be based on their lived experience, but they can (and should) use writing assignments as a chance to generate and articulate some new understanding about themselves, their families, their impact on others, and their roles in communities.

Regardless of the instructor's motivation, personal writing is widely taught in composition courses, and it not only provides a good departure point, but also illustrates stylistic techniques students can use in other types of writing to invigorate their prose. A simple anecdote can provide the framework for a coherent and compelling paper on something seemingly unrelated; a childhood memory can spark an image that unifies an argument; a reference to an event can illustrate a point in a paper that might otherwise lack a personal connection. Furthermore, with the advent of postmodern approaches in both teaching and scholarship, autobiographical writing is breaking down the barriers of more formal writing styles accepting techniques that have previously been considered inappropriate. This chapter provides the opportunity to introduce students to some of these strategies.

Confronting the Issues

Option 1: Constructing Contexts

Before you assign any of the reading selections in this chapter, ask your students to make a chronological list of what they consider the major milestones of their lives—the "bests," "worsts," "firsts," "lasts," and "onlys." Then ask students to conduct an interview (in person or by telephone) with one of their parents. In this interview, parents should do just what the students have previously done: They should trace their children's lives from birth to the present, identifying significant milestones.

After the interview, each student should compare the two lists and write a paragraph or two accounting for any discrepancies between them. In class, encourage students to generalize about the differences between the way they see their own lives and the way their parents see them. For example, are their parents more or less likely than the students to remember negative milestones? Which years do parents (and students) recall most vividly? Most favorably? Are there any events that students see as positive and parents as negative, or visa versa?

Finally, introduce to your class the idea that many of this chapter's selections depend on memory and on subjective views of parents or other family members. Students will most likely realize by now that such readings cannot present a "true" or "complete" picture of a family member's life or of relationships within families because there is no single, complete "truth." As they read the selections in this chapter, then, they should be concerned not just with facts but with subtleties: motivations, emotional reactions, and differing or changing points of view.

Option 2: Community Involvement

Many of the writers in this chapter describe childhood traumas or aspects of their dysfunctional families. Sometimes their first year of college is the first time students become aware that their own families' shortcomings are not unique, so these readings may be quite surprising and cathartic for some of them. Conversely, other students may be surprised and irritated that so many readings deal with family problems. You can use these tensions as an opportunity to have students research a bit about the outreach services available in the community surrounding your institution, and they can complete the assignment by writing a handbook for incoming students outlining social services available in the college community. First, students can generate an initial list of family-centered problems incoming students might have; this list may grow as they read the pieces in this chapter. Next, you can bring in a local telephone book, a list of services provided by your institution, any pamphlets from nearby public health services, and any other reference sources you can identify. Students too should try to locate such services. Working in groups, students can find out more information about the various resources available to students: hours, fees (if any), hot line telephone numbers, places and times for support groups, types of aid offered, and so on. Finally, they can put it all together in a form that can be distributed to other incoming students (or to members of the community) as a public service.

Option 3: Cultural Critique

If you like to incorporate cultural critique into your class, this is a perfect chapter for examining the myths surrounding the "American Family." As a class, have students generate cliches and phrases that typically describe the American family: "Home is where the heart is," "There's no place like home," "Blood is thicker than water," "Spare the rod and spoil the child," "The family that prays together stays together." In groups, then, students can offer a positive translation of what each saying is supposed to connote. If necessary, supplement these discussions with segments of Dan or Marilyn Qualye's "family values" speech at the 1992 Republican Convention; family-oriented cartoons like *Family Circus, For Better or Worse, or Hi and Lois*; or snippets from TV shows such as *Frazier, The Simpsons,* and so on.

Next, students can offer the other side of the story to many of these clichés and images without any help from the instructor. Sometimes the contradictions are readily apparent. Others, however, will be more difficult, and you may want to try some other strategies. For instance ask students to identify the assumptions implicit in each phrases: *socioeconomic, race, level of education, religious affiliation, gender roles, sexual orientation,* and so on. Illustrate the type of family

4

each of these phrase describes; then illustrate the types of "families" that are excluded. Many of those families are described in the readings throughout this chapter.

Option 4: Feature Film

The film *A River Runs Through It* deals with many of the issues that are raised in the readings of this chapter: rebellion, coming of age, alcoholism, competition, fragmentation, and death. For a composition class, this is a particularly nice starting point, because the narrator is a writer, and writing plays a central role in several scenes. If time permits, begin the unit with a screening of the movie, and use it to compare and contrast with the readings you select from the chapter. Other possibilities include *Ordinary People, The Great Santini, Housekeeping,* and *Running on Empty.*

Teaching "Two Perspectives"

As students read the following poems, ask them to notice the difference between the two narrative perspectives; one is written in the past tense and the other in the present tense. Discuss the powerful emotions children can feel toward their parents: guilt, admiration, and a desire to please.

"THOSE WINTER SUNDAYS," ROBERT HAYDEN

For Openers

"What did I know, what did I know?" What does the poet know now that he did not know as a child?

Collaborative Activity

Ask your students to read Raymond Carver's poem "Photograph of My Father in His Twenty-Second Year" (included in his essay "My Father's Life," which appears earlier in this chapter). Then assign students, working in groups, to identify similarities and differences in the two speakers' attitudes toward their fathers.

Writer's Options

Rewrite "Those Winter Sundays" in the form of a eulogy to be delivered by the son at his father's funeral. You may use language from the poem, but you should also blend in invented details that will characterize the father and present his struggle in specific terms.

"DIGGING," SEAMUS HEANEY

For Openers

Discuss the expectations established by Heaney's phrase "snug as a gun" in the first stanza. How does that metaphor shape the rest of the poem, as Heaney drifts from generation to generation? What does the gun image say about Heaney's self-identity?

Teaching Strategy

Heaney draws a connection between the tools of his father and his grandfather, and the tools he uses as a writer. Do you follow his reasoning? Do you believe that his pen is his spade? Can you write a counter-example that would challenge the power of writing?

Collaborative Activity

In groups, students can generate the images Heaney creates for one of the generations. What is each man's work? What images do they associate with that work? What does "digging" mean in the context of that work? What values are associated with that kind of work? What class identity does each have? Have students put their group's responses on the blackboard and ask the class to reach a consensus about how each generation's "key terms" contrast with the other two.

Writer's Options

Is there a common tradition in your family, or a vocation that previous generations share? Identify a common image you can use to tie that tradition in with your own experience. Heaney uses "digging" for all three; what image unifies your pursuits with those of the rest of your family?

Multimedia Resources

For a more extended look at Irish agriculture, show students parts of the film *The Field*. The scenery alone will provide some visual context for Heaney's poem. In addition, contrast struggles to cultivate the land and fighting with the development plans of the younger generation. Students can explain how "Digging" comments on that tension explored in the movie.

1. The memories show both fathers as hard workers and hard men with "cracked hands" and "coarse boots." However, the household in Hayden's poem shares a "chronic anger" and indifference, a climate not found in Heaney's poem.

2. The father has warmed the house and polished his son's shoes. But the son seems to have feared the "chronic anger" of his father's pent-up and unexpressed frustrations.

3. Heaney writes of his father and grandfather with immense respect, treating their crafts of digging as he treats his own writing. In this poem, he questions whether or not he can sustain the authority, precision, and skill with which those two men worked. Hayden does not share a similar connection with his father.

4. Both of the poets are looking to past generations for lessons they can use in their current lives. Each finds new respect for the traits they might have previously misunderstood or overlooked.

5. The speaker in Hayden's poem may have inherited the ability to serve his family silently, without reward, but with honor and satisfaction. Heaney's speaker has inherited the ability to dig beneath the surface of words to look for hidden fruits and meanings.

6. Answers will vary.

Using Specific Readings

"ONE LAST TIME," GARY SOTO

For Openers

Based on what Soto tells us, is this really the "one last time"?

Teaching Strategies

Discuss the following with your students:

1. In a set of autobiographical notes, "Comments addressed to Juan Rodriguez, May 1977," Soto writes: "I write because there is pain in my life, our family, and those living in the San Joaquin Valley. I write because those I work and live among can't write. I only have to think of the black factory worker I worked with in LA, or that toothless farm laborer I hoed beside in the fields outside Fresno… they're everything." Ask your students to comment on this statement as it relates to what they have read. Does it provide further insight into the emotional quality of the piece?

2. Ask students to list the "best" and "worst" jobs they can imagine. Then ask them to establish specific criteria for what makes a job "good" or "bad." Discuss the implications of each type of work, paying particular attention to what values are generally held by those in the "best" jobs and those in the "worst." Analyze the assumptions students have about each type of job, looking especially at race, class, gender, required education level, religious affiliation, and sexual orientation. What assumptions are implicit in the way we (and our students) view these kinds of work? Who has access to the "best" jobs? Who does not?

Collaborative Activity

Soto emphasizes his motivation for dressing in a particular way. In small groups, pick one type of clothing he mentions and describe its connotations. Every group should analyze a different article of clothing. Students can report on their findings, looking for trends throughout the essay. Are there odd contradictions? If so, what do they indicate about Soto?

Writer's Options

Write about the most difficult job you ever had. In what ways were its challenges similar to or different from those of Soto's job? What motivated you to continue to work despite these difficulties?

Multimedia Resources

1. Woody Guthrie wrote some songs that evoked the plight of mid twentieth-century migrant workers. One in particular, "Deportee," eulogizes "nameless" Mexican workers whose plane crashed after they were expelled from the United States.

Guthrie gives them back their names and their dignity in this period song. Play it in class to provide some historical context for Soto's descriptions of his migrant farming experience.

2. Bring in a videotape of the "Heard it through the Grapevine" California Raisin Claymation commercial, and encourage students to critique the ad based on the perspective they have gained from reading about the raisin industry in "One Last Time." To whom does the ad appeal? To what values does it appeal? What are the unspoken assumptions in the ad? After their critique, let students know that this ad won several awards and spawned not only sequel commercials, but an entire TV special and scores of stuffed animals and other merchandise.

Suggested Answers for Responding to Reading Questions

1. Work enables Soto to understand his mother's experiences and hardships — and her warnings. He realizes how hard it is for her to earn money and the reasons for her "smoky view of the future." Soto's experiences are typical: He endures the boredom and hardship of work, but he would prefer to be relieved of such tedium. He has an understanding of his own needs.

2. Soto probably sees in the film the same kind of harsh physical labor for poor remuneration that he experienced in his own childhood.

3. The adult Soto understands his boyhood pride. He knows that children are easily ashamed in front of their peers. As an adult, Soto sympathizes with his boyhood self. The work is no more appealing now than it was then.

Additional Questions for Responding to Reading

1. What triggers Soto's memory? What is the specific association between the Indians in the film and the people Soto remembers?

2. What warning does Soto's mother repeatedly offer? Have you received warnings from your parents that you took more seriously when you matured?

3. What things seem to be valued by the youth Soto describes? What things does he himself seem to have thought were worth working for? Are these typical teenage values.?

9

4. What does Soto mean in paragraph 10 by "boredom was a terror almost as awful as the work itself." Is he exaggerating?

"ONCE MORE TO THE LAKE," E.B. WHITE

For Openers

Remind students that White is looking back over many years. Given this distance, and given his obvious affection for his father, do students question the accuracy of White's memoir? Do they perhaps see his view of his childhood (and of his relationship with his father) as unrealistically, even impossibly, idyllic? Judging from their own experiences, what kind of details or incidents do they suspect White may be omitting?

Teaching Strategy

Part of the vividness of White's essay derives from the clarity with which he evokes particular details. Have your students make note of especially strong images, and ask them to talk about what makes these images so evocative.

Collaborative Activity

Ask students to work in groups to list the specific images White uses to describe the lake. Assign one group to list sounds associated with the spot, two or three groups to identify visual images, and one group to compile a list of tactile or other images.

Writer's Options

Write about an experience you had that seemed to erase (or reinforce) the differences between generations.

Multimedia Resources

Bring in some fine art prints or slides of nature scenes and compare them with the feelings that White evokes. American Renaissance painters would work, as would members of the Hudson River School.

Suggested Answers for Responding to Reading Questions

1. White claims that, in all significant respects, the lake has not changed. The essay focuses on his return to the lake as an adult

and on the way the passage of time seems to have been a mirage.

2. While the essay is heavily based on the description of a particular place, the essay is about the mysterious nature of time and the relationship between fathers and sons.

3. Although one might need to be a parent to share White's particular reexperiencing of himself in his son, many of us recall the youthful sense that things will never change. As we mature, we realize that summer is not "without end" and that there have indeed been (and will continue to be) profound changes in our lives. Other answers will vary.

Additional Questions for Responding to Reading

1. In what ways is this an essay about the links between generations? In what ways is it also about the gaps between generations?

2. What aspects of the lake seem disappointing when White revisits them?

3. White's essay takes us backward and forward in time. What stylistic techniques does he use to signal shifts in time to readers?

"NO NAME WOMAN," MAXINE HONG KINGSTON

For Openers

Students are likely to be repulsed by the brutality and cruelty described and may even object to having had to read it. You might want to open with a discussion of whether any of the unpleasant details could be omitted. Ask students what effect such omissions would have on the essay. Would they, for example, make the essay less forceful?

Teaching Strategies

Discuss the following with your students:

1. Kingston's story weaves the "real" and the supernatural together just as it intertwines the Chinese past and Kingston's memories of her American girlhood. How does Kingston accomplish this interweaving of "fact" and fantasy? What

transitional devices does she use to link them? What tone and mood are established through this effect? Kingston, as the narrator, tries to fill in the gaps in the stories about her aunt in order to make a complete image. Locate and discuss the different interpretations of *Aunt* throughout this piece, and show how that definition shifts as the narrative goes on.

2. There are certain values that Kingston does not share with members of her parents' generation. How would you characterize those values? Are there times in your life you can tell you differ from your family or community? Do you think those disagreements are because of the generation gap, or is it something else?

Collaborative Activity

Assign students working in groups, to brainstorm about who was most guilty of causing pain to Kingston's aunt and who, if anyone, could have saved her reputation (or her life).

Writer's Options

1. Write a fictionalized account of a person who is a skeleton in your family's closet. This person need not be somebody close to you. Invent aspects of his or her life story that led to the person's questionable status in your family. Does the process of writing additional (even made-up) information about this person alter your perceptions of him or her? How?

2. Kingston claims to be surrounded by an "invisible world." Are you conscious of having an invisible world of ancestors? Write an essay in which you describe your invisible world or the reasons why you might not have one.

3. Write a case study of Kingston's aunt from the point of view of a Western observer. Present her situation, and offer recommendations about how her problems could have been alleviated. Try to use an objective, "scientific" tone.

Suggested Answers for Responding to Reading Questions

1. Kingston's "facts" are adequate for the story that she chooses to tell. The story is constructed out of Kingston's knowledge of her

people, her suppositions based on this knowledge, and her intuitive sense of her aunt's experiences.

2. The author's relationship to the unknown aunt has profoundly affected her growing up. The story has made Kingston believe that "sex was unspeakable," that words were powerful, and that fathers were frail. The mystery surrounding the aunt has haunted Kingston and seems to be connected to her own conflicts with maturity and with defining her Chinese-American identity. Because so little is known of this aunt, she may have served as an emotional projection of Kingston's own inner conflicts.

3. Answers will vary.

Additional Questions for Responding to Reading

1. Kingston's mother used the aunt's story as a warning to her daughters. Does the lesson appear to have had the desired effect? Does Kingston imply that it was a valuable lesson? Is Kingston's interpretation of the story different from her mother's? Explain.

2. In what ways is this essay about not speaking? What does this silence suggest about the power that is invested in words?

3. Does Kingston appear to judge harshly any of the parties involved in the aunt's story?

"THE WAY TO RAINY MOUNTAIN," N. SCOTT MOMADAY

For Openers

Remind students that Momaday is a poet as well as a writer of fiction and essays. In what respects is this essay "poetic"?

Teaching Strategies

1. Consider the strategies Momaday uses to make his grandmother a mythical, larger-than-life figure. Do these strategies make her more or less human? Why do you think he uses these strategies?

2. Momaday is connected to his people's history through his grandmother. What values do you have because of your ethnic or cultural heritage? What is that heritage? Who serves as your

connection to that heritage? How do you maintain your connection with that person?

Collaborative Activity

Ask students, working in groups, to identify instances in which Momaday describes his sense of awe in viewing certain features of the landscape. Then assign each group to experiment with rewriting a different one of these descriptive passages in plainer, more objective language. As a class, compare the two versions of each description and discuss the advantages and disadvantages of each.

Writer's Options

1. Momaday's grandmother sees "more perfectly in the mind's eye" (paragraph 5). Write about a place or event that you did not experience yourself, yet you envision vividly. Write as much descriptive detail as you can. What is the significance of this event or these images to you?

2. Through his grandmother, Momaday is connected to his people's history. Write an essay in which you discuss how your grandparents connect you to your ethnic and cultural heritage.

Suggested Answers for Responding to Reading Questions

1. The memories Momaday recounts seem to make the grandmother a symbol of the once powerful, but now dying culture. Her life is emblematic of the culture and history of the Kiowa people.

2. Whereas we do not know Momaday's grandmother as an individual, her historical and cultural background enables us to understand more about the Kiowa people. We come to know something of their pleasures, hardships, struggles, and defeats.

3. Momaday's journey is as much an imaginative, or a spiritual, excursion as it is a physical one. In writing about his grandmother, Momaday reasserts his own identity.

1. What do Momaday's images of enclosure suggest about the Kiowa people? What aspects of life and nature do the people seem to cherish?

2. Momaday writes that his grandmother "must have know from birth the affliction of defeat." What does he mean? Does he expect readers to take this observation literally?

"BEAUTY:WHEN THE OTHER DANCER IS THE SELF,"

ALICE WALKER

For Openers

Walker's accident transformed her life in many ways. How might her fear of blindness have affected her development as a writer? (You might mention to your students that some great writers, such as Homer and Milton, found that blindness paradoxically led them to new insights.)

Teaching Strategy

Walker frequently understates emotional reactions. Ask your students to identify instances in which they believe she is understating such reactions. Why do they believe she uses this flat, relatively unemotional tone to recount particularly painful experiences?

Collaborative Activity

Ask students, working in groups, to experiment with recasting Walker's narrative into a screenplay. Assign a different section of the essay to each group, and have them create appropriate dialogue, descriptions of costumes and scenery, and so forth.

Writer's Options

1. Walker uses repetition to unify and underscore her insights: "now that I've..., now that I've..., now that I've...," in paragraph 32; "you did not change" in the first two-thirds of the text; and "I remember" in the final third of the text. Try this technique

yourself; pick a starting point for your own recollections and repeat the phrase to introduce each new idea.

2. Walker's accident affected her development in unforeseen ways. Write about an experience whose full consequences you did not comprehend until some time later.

Suggested Answers for Responding to Reading Questions

1. Present tense conveys the action and tension of the story more immediately to the reader. It also raises sympathy in the reader more effectively.

2. Walker repeats the phrase several times to emphasize the gravity of this incident and the irony of the fact that she did change. Her repetition makes clear the degree to which her family was unaware of her childhood trauma.

3. The adult writer seems not to blame anyone for her childhood problems, though she may blame the situation that allowed it to happen. Clearly race and gender were significant factors in the outcome of the accident: She would have reached the doctor sooner had her father not been black, her brothers would not have had BB guns had they not been male, and she might not have felt there was such a premium on beauty had she not been female.

Additional Questions for Responding to Reading

1. What is the effect of Walker's recounting her memories in the present tense? What does this form of narration suggest about her ability to resolve her childhood pain?

2. Notice those words around which Walker has placed quotations marks. What is suggested by the following: "beautiful," "on the ranch," not "real" guns, "cute," and "accident"?

3. What strategies does Walker use to break her essay into sections?   Why does she do so?

"MY FATHER'S LIFE," RAYMOND CARVER

For Openers

What, besides the father's life, is the focus of this essay?

16

Teaching Strategies

1. Is the father presented as an individual or a symbol? Although the essay's last line suggests that to Carver he is an individual, students may see him as something else as well. If he is a symbol, what does he represent?

2. Students will need to establish the distinction between the mature narrator's attitude toward his father and his probable feelings as a boy. What does the narrator understand that he seems not to have understood before? You may want to try taking some guesses as to Carver's possible motives in writing the piece.

3. Carver expressly describes his process for and reasoning behind choosing "October" as the setting in the poem, despite the actual month having been June. Use these final paragraphs as a springboard for discussing word choice with your students, and discuss the implications of changing other pieces of the setting.

Collaborative Activity

Ask students to work in groups to analyze the way in which Carver conveys an emotional attitude by showing rather than telling. Assign a different section of the essay to each group. In class discussion, consider these questions: Why does Carver choose to include the sorts of details he does? What is the cumulative effect of these details?

Writer's Options

1. Both Carver and Walker include a poem in the middle of their essays. Include in your journal any poem you may have written in response to an event, or recopy a poem you like and explain its significance to you.

2. In paragraph 6 Carver describes his father's reaction to Franklin D. Roosevelt's visit to the Hoover Dam. As he observes, visiting dignitaries on public relations missions often praise the apparent good of a project without mentioning (or even knowing) the downside. Can you think of a time when this happened in your own neighborhood or school? What were those unmentioned negative aspects of the project or proposal?

Multimedia Resources

If your students are intrigued by the pictures Carver creates of dysfunctional life, bring in clips from Robert Altman's *Short Cuts*, a bleak film based on vignettes from Carver stories. In class, follow those threads that might interest your students most.

Suggested Answers for Responding to Reading Questions

1. Carver's essay is a clear demonstration of the precept that literature should show and not tell. By including these seemingly mundane details about his father's life, Carver builds a believable portrait of the man.

2. Answers will vary. The son never criticizes the father, nor does he imagine himself as an equal to his father, as another man who is vulnerable to life. The poem allows the mystery of the father to be displayed in ways that prose does not.

3. Carver uses the poetic form to express an emotional attitude toward his father from his mature perspective. In the poem he comments on his father's youth, confusion, and vulnerability. Carver conveys a sense that, in spite of whatever conflicts may have arisen between the two "Raymonds," the poet finds himself finally unable to criticize his father's "weakness." The poet seems to have come to a similar understanding of life's hardships.

Additional Questions for Responding to Reading

1. What might be the significance of the names and dates that are so precisely recounted to us in the first section? How does Carver's emphasis on these details lead to an understanding of his theme or themes?

2. What definitions of *work* does Carver present? How does his view of work seem to differ from his father's? Which definition is closer to your own?

3. Analyze paragraph 4, in which Carver's mother discusses her relationship to her husband. Can you paraphrase the complex of emotions? What does she seem to have felt for the man to whom she was married?

4. What aspects of Carver's father's life surprise you? How does this lack of predictability seem to have affected the narrator?

Focus: Is Divorce Destroying the Family?

The following three essays challenge a stereotypic notion of "family," showing that the spirit of family exists beyond the bond of the parents' marriage. Then ask them to list different family combinations of adults and children. Then ask them to reflect on whether there is a typical '90s family.

"THE PERFECT FAMILY," ALICE HOFFMAN

For Openers

Hoffman writes, "We ourselves did not dare to be different. In the safety we created, we became trapped." Begin class with a freewriting period to allow students to respond to Hoffman's statement. Follow with a discussion on how such isolation can occur as a result of other aspects in one's life, such as dress or religious practices.

Teaching Strategy

The word *perfect* is a pivot point in this essay. Discuss the different "perfect" situations that are described, directly and indirectly, in this essay. How can expecting perfection actually work against a person (e.g. the perfect party, relationship, dinner, face, body, game, teacher, essay, son, daughter)?

Collaborative Activity

Compare "chore experiences" among the group members. See if there are differences in how chores are assigned in families with daughters, sons, or mixed genders--or in single or two-parent or multigenerational families. Consider simple chores, such as taking out the garbage or mowing the grass. Do television programs reinforce chore assignments?

1. This essay claims, "People stayed married forever back then, and roses grew by the front door." Write an essay that idealizes your own childhood. How might someone describe or define 1990 families in idyllic terms?

2. Write a response to Hoffman's essay from the view of a single mother or a parent in a two-income family.

3. Beyond the clean matching curtains and furniture, the word *home* means something more specific. Using examples from real families, fictional families, and families in the media, write a descriptive definition of what *home* means.

4. Read Nikki Giovanni's poem "Nikki Rosa." How does this poem relate to this essay?

Multimedia Resources

Check a library or a video store for copies of television programs cited in this essay. While viewing the shows, ask students to list the characters' actions that show stereotypic role playing. Notice how nicknames are used. For example, in "Father Knows Best," the oldest daughter is called "Princess" and the youngest is called "Kitten." The son is called "Bud." What effect do nicknames have on the characters and on your perception of that character? What characters do not have nicknames? Is this significant? Father has his own study. Mother is in the kitchen. How are character locations significant?

Suggested Answers for Responding to Reading Questions

1. The Hoffman family was a single-parent family as a result of divorce. Today, divorce is much more common, yet children of divorced parents may still feel "different." Families today seek sameness in other ways: in styles of dress, places to vacation, choice of cars, the kinds of technological items owned. At one time, suburbs were very homogeneous; at least on the surface, families seemed alike.

2. Today, the concept of family can include single-parent families, multigenerational families, multiracial families, families with gay or lesbian parents, and families with various disabilities.

Responsibilities are shared differently from family to family as well. Since the 1950s, the family has found its model in European patriarchal society; the perfect family would be a two parent home where the father worked outside the home and the mother was the housewife. Today, other options are available for men and women--including a return to the two-parent (one working parent and one stay-at-home-parent) household.

3. The simple rules of the suburb meant that the father left for work in the morning and returned at dinner time. Mother cooked, cleaned, and managed the children's daily affairs. The children went to school and came home to play and do homework and certain chores. In the Hoffman household, more than just family members came in and out of the house. The "perfect" suburban families, like those with whom Hoffman's mother worked, had some serious problems; in other words, they *weren't* "perfect."

Additional Questions for Responding to Reading

1. Review the essay for words that mark time, such as "when I was growing up" and "a few years later." Underline or highlight these words and phrases. Consider how these time markers add cohesion and coherence to the essay.

2. Study how the essay shifts its attention from the mother to the child and then to the narrator, now a parent. How does Hoffman achieve these the changes in the perspectives?

"STONE SOUP," BARBARA KINGSOLVER

For Openers

Read the children's story "Stone Soup." Discuss how the soup is made differently at different times. What is wonderful and dangerous about eating stone soup for supper?

Teaching Strategies

1. Spend some time helping students decode the text by listing the negative phrases, similes, and word choice used to create sympathy for the writer's position. Ask why the term "failing" has such reverberations. With what is it associated?

2. Pay attention to authorities cited. Ask why they are convincing or weak.

3. Read the Sharon Olds poem "Living in Sin." What are the ambiguities in the poem?

Collaborative Activity

Bring in magazine ads or catalogue shots that portray families. How is the concept of family depicted? How do these illustrations resemble the "Paper Doll Family"?

Writer's Options

1. Argue that fairy-tale families are *not* the most happy of families. Set up criteria for what makes a happy family and refer to specific tales.
2. How would Hoffman's "Perfect Family" respond to Kingsolver's "Stone Soup"? Explain the difference between their arguments.

Multimedia Resources

1. Rent the movie *The Bird Cage* with Robin Williams. Discuss how the concept of family shifts and develops in the film.
2. How are families defined or described by TV talk shows or radio "help" programs? What are audiences looking for in these programs?

Suggested Answers for Responding to Reading Questions

1. The family certainly has many pieces in it. Is it broken? Answers will vary.
2. The family of dolls represents perfect combinations, except they are plastic, come in boxes, and are bought and sold. Kingsolver feels antipathetic toward this life style.
3. These terms suggest that the opposite sort of family, the whole and healthy family, suffers no controversy, argument, or altercation. Students will have their own opinions, and while "unbroken" families can be dysfunctional units, the statistics are being used here to show that the more stable home life is a positive influence helping its member become more successful.

1. Review the essay looking for figurative language, for example, "this narrow view is so pickled." How does figurative language add texture to the essay?

2. Several sentences begin with the personal pronoun "I." Is this a strength or a weakness in the writing?

"SPLITTING UP," JOSEPH ADELSON

For Openers

The author uses the term *institution* in reference to marriage. Discuss both the positive and negative implications of that term. What might the "institution of marriage" mean to his parents' generation? How is this meaning different from what marriage means today?

Teaching Strategies

Consider other social practices that may have at one time been considered "problems, but now are accepted." Focus this response on Adelson's passage: "In the case of divorce, Popenoe suggests, one reason we have tended to ignore or shy away from the magnitude of the problem is that so many of us are already divorced, and we do not like to think of our individual experiences as symptomatic of a troublesome *social* phenomenon. Another reason is undoubtedly the sense that to begin thinking of divorce in this way would in any event be an exercise in futility; the practice has now become just too deeply woven into the fabric of our social life." You might consider out-of-wedlock pregnancies, high school dropout rates, and the like.

Collaborative Activities

1. In "Stone Soup," Kingsolver says, "But there's a current in the air with ferocious moral force that finds its way even into political campaigns, claiming there is only one right way to do it, the Way It Has Always Been." Ask groups to discuss how Adelson might respond to Kingsolver.

2. "Splitting Up" begins with negative vocabulary to describe divorce, as do the preceding two essays. Study how the essay progresses from similar openings to divergent claims. Ask students to work in groups and then compare their conclusions.

Writer's Options

1. Find current research on the effect of divorce on children that counters Adelson's evidence. Use the different studies to help you reach a conclusion in an essay of your own.

2. Acting as a moderator for a discussion on a make-believe talk show or an Internet chat room, create a dialogue between Adelson and Kingsolver or Hoffman. As moderator, what conclusions can you draw?

Multimedia Resources

1. Ask students to list titles of popular love songs, from country to show tunes to operettas. Are the songs primarily about loss in a relationship, or are they celebration of commitment?

2. Consider some popular television sitcoms. What is the family arrangement on these shows?

Suggested Answers for Responding to Reading Questions

1. Experts explain that when what previously would be considered socially unacceptable behavior becomes prevalent in a society, then, for fear of raising the specter of a huge problem, society shrugs it off and accepts it by weaving it "into the fabric of our social life."

2. Answers will vary according to individual experiences.

3. While Kingsolver sees no mystique in the two-parent home, but trusts that deep caring will create a successful family, Adelson holds that divorce is a almost always detrimental to the parties involved, especially the children. The differences may be caused by their personal histories or by their different professions.

Additional Questions for Responding to Reading

1. Highlight all passages that cite authorities or researched data. Read the sentences before and after the highlighted passages to see how the author incorporated and commented on source material.

2. Consider how the last four paragraphs function individually and collectively to bring the essay to closure.

CHAPTER 2

ISSUES IN EDUCATION

Setting Up the Unit: Combining "Hard Facts" with "Soft Stories" for Persuasive Balance

Education is a popular topic in composition classes because it is a subject with which students have ample experience. Students will have only limited knowledge, however, of what actually comprises "education" in situations far removed from their own. This chapter provides some alternative pictures: some from students, some from teachers, some from researchers, some in favor of curriculum changes, some fighting to maintain a standardization they feel equalizes; some questioning whether any change can be enough; some questioning whether the changes currently underway are ill-advised. The essays in this chapter paint pictures of education across the socioeconomic scale and across age groups. With a topic like education, students can use their own knowledge, assumptions, and perceptions about what education can (and cannot) provide; these readings will offer more perspectives, more depth to your discussions, and more points of departure for your students' papers.

Many, if not all, of the essays in this chapter share a structure that balances personal (or observed) anecdotes with the statistics and demographic information. In all cases, the author takes a position on the issues surrounding education; each writer, however, gets his or her point across with very different emphases, very different styles, and different ratios of "hard" data to "soft" anecdotes. All writers, however, use both. As your students work their way through the assigned readings, have them keep a running total (perhaps in their journals) of the ways in which each writer balances personal writing with what the writer presents as *objective truth*.

In addition to keeping track of the "objective"/"subjective" mix, students can also chart how and when each type of writing occurs. When do most hard facts appear? And when do they seem to be most effective? Likewise, when do personal anecdotes or stories seem to tell more than the numbers? When do they appear most frequently? After gathering this information, students can then attempt to replicate this balance of statistics and anecdotes in their own papers.

Clearly, teaching these writing strategies can also *raise* questions of what constitutes objectivity and truth. You can encourage students to test the writers' assertions against their own experience and discuss their insights with the entire class. Likewise, they can do some informal "field research" by talking to their friends and colleagues at other institutions, thus gaining a broader perspective than might be possible within your institution alone. The selections in this chapter take a wide range of contentious points of view, and thus provide something for every student to contemplate.

Confronting the Issues

Option 1: Constructing Contexts

Many students are not aware that a debate is taking place that will decide the direction of their education. On one side are those who claim that the standard core curriculum at most American schools, with its stress on Western values, does not reflect the reality of the United States today. They point out that the 1990 census shows the United States to be an increasingly diverse country, made up not just of white males of Northern European descent, but also of women; African, Hispanic, and Asian-Americans; and people of Eastern European descent. Most experts agree that by 2005, most students will be people of color. For this reason, the argument goes, the curriculum must be changed if it is to be relevant to students' lives and to connect their unique experiences to the learning process. In a high school with a diverse ethnic population, for example, the study of American history could treat topics such as Asian immigration, the role of African-American soldiers in the Civil War, and the effect of Jefferson's ideal vision of America on colonial women. According to Henry Louis Gates, Jr., W.E.B. DuBois professor of the humanities at Harvard, "It's only when we are free to explore the complexities of our hyphenated culture that we can discover what a genuinely American culture may look like."

26

On the other side of the debate are those who reject the call for multicultural education, which they believe would ultimately fragment the curriculum into a multitude of separate ethnic enclaves. At its extreme, they charge, multicultural education rejects the values and beliefs of Western civilization. They maintain that this heritage has created a society that, however flawed, has influenced writers and thinkers of many backgrounds and given the world the idea of a government that champions the dignity, independence, and freedom of the individual. For this reason, says Donald Kagan, dean of Yale College, "It is necessary to place Western civilization and the culture to which it has given rise at the center of our studies, and we fail to do so at the peril of our students, country, and the hopes for a democratic, liberal society."

After you explain the two positions outlined above, ask students what they think of the idea of a multicultural curriculum. Have they been exposed to such a curriculum? What are its advantages and disadvantages? Is there a way of bridging the gap between the two positions stated above? Do they see any value in studying history, literature, or even sciences in this way?

After this general discussion, divide the class into groups and ask each group to reexamine one of the reading selections in this text that the class has read. Their assignment is to evaluate the essay, poem, or story in light of its value to their college's student body, considering subject matter, style, date of publication, author's background, and any other criteria they like.

In further class discussion, debate the usefulness and educational value of the selections students have identified as particularly meaningful to—or particularly remote from—their experiences. Why should today's students read each selection? In short, does education mean reinforcing or challenging ways of thinking and learning about issues and ideas? Guide the discussion toward a development of a definition of "education."

Option 2: Community Involvement

What issues face the schools in your area? Read through the community newspaper to find an issue in a local school that pertains to a topic in one of the chapter's readings. For their project in this chapter, students can pose possible solutions to that problem, ideally by volunteering or setting up a program, or by doing research to find out what sorts of support systems might already be in place. One example: At one large midwestern university, the football team sponsors a "Gentle Giant" program through which members of the football team tutor at a local elementary school.

Option 3: Cultural Critique

Begin by generating as a class commonly agreed-upon "truths" about public education in America: It is free to all, it produces better citizens, it teaches the "basics," it guarantees success. Next, choose an educational setting that your students know well, and ask them to chart both the "hidden" and the "explicit" curricula they identify; charts like these can expose the differences between what schools say they teach, and what they actually do teach. As your class reads through this chapter, continue to map where each piece fits into the chart: What are the constraints in each situation? What is the ideal to which it strives? Describe the distance between the deal and the actual in each of these settings.

Option 4: Feature Film

The first several essays in this chapter are about underprivileged schools and the hardships they face. Begin the chapter with a viewing of *Stand and Deliver*. Based on a real math teacher and his students, this is a useful film because it both upholds and critiques the ideals of the American educational system. In the home video version, the actor portraying the teacher describes what an honor it is to stand in this man's shoes, and he gives an impromptu speech about the value of an education. By all means, show this part in addition to the film itself. The film's story uncovers the contradictions that poor, urban teachers face: They are caught between wanting to challenge students and not wanting to demoralize them; they are caught between a love of teaching and the temptation of a higher-paying job; they are called upon to work above and beyond the standard school hours in order to actually make a tangible difference. *Stand and Deliver* also depicts the impediments students have to face, including the Educational Testing Service's belief that so many students could have passed the AP Calculus exam only by cheating. At the end, however, the tag lines only tell us how many students have passed the exam in subsequent years; they do not tell us how their lives were changed — if at all — by this experience.

Teaching "Two Perspectives"

Before reading Twain and Whitman, ask students to reflect on the difference between arts and sciences. How is objective interpretation different from the subjective? Are they necessarily opposed to each other? Discuss how reason and imagination are related.

"READING THE RIVER," MARK TWAIN

For Openers

Most students will come to this essay assuming that education is always beneficial. For this reason, Twain's ideas will seem foreign to them. A few minutes spent discussing what is lost as well as what is gained through education may be in order. Introduce the loss of innocence motif with the story of the fall from the Garden of Eden. Ask students whether they believe that knowledge inevitably leads to disillusionment and a loss of innocence.

Teaching Strategy

Ask students for examples of how their views of a place, an object, or a person changed as they grew older. What caused the change?

Collaborative Activity

Divide the class into groups and distribute to each group a copy of a fairy tale such as "Little Red Riding Hood" or "Where the Wild Things Are." Ask each group make two lists, one of details an innocent, unsophisticated reader (such as a child) would notice, and one of details an educated adult would see. In class discussion, compare each group's lists. What can be inferred about the two ways of approaching a story?

Writer's Options

1. Write an essay in which you compare how things looked to you the first day you saw your college campus and how they look to you now.

2. Write about an incident that taught you that things were not what you thought they were. Did you feel more mature or simply betrayed?

"WHEN I HEARD THE LEARN'D ASTRONOMER," WALT WHITMAN

For Openers

Discuss the title, especially the choice of the word *learn'd* to describe the astronomer. What connotations does the word have? In what ways is he learned? In what sense could he be considered ignorant?

Teaching Strategy

Compare this poem to Twain's "Reading the River" (page 98). In many ways both Twain and Whitman express the same, or at least similar, sentiments toward education. Ask students if they think one writer is more hostile to education than the other. Then ask them to explain the differences between their attitudes.

Collaborative Activities

1. Assign groups to look for examples of alliteration, varying line lengths, lists, and repetition. What effect is produced by each strategy?
2. Have each group of students summarize each segment of the poem in a single sentence. Compare the contrasting ideas as they are expressed in prose and in the poem.

Writer's Options

Describe how education changed your view of something. Be specific when describing your original and your changed viewpoints.

Two Perspectives: Suggested Answers for Responding to Reading Questions

1. Twain seems to prefer the innocent view, since it retains the mystery, beauty, and excitement of nature. Twain is, however, a realist. Even though he appreciates the romantic view of the river, he realizes that a realistic view is absolutely necessary if he is to navigate the river safely.

2. His attitude could be called resigned. Twain knows that education inevitably results in a loss of innocence, but he also knows that this loss is necessary.

3. The first segment expresses the scientific approach; the second, the romantic, imaginative approach.

4. The speaker devalues formal education and emphasizes the possibilities of spontaneous, individual responses to life.

5. The points could be conveyed through narrative or by using statistics. Any other genre could handle the issue and concepts of education.

6. Answers will vary. Most will agree that education often strips the world of its mystery, and yet some may think of situations in which the mystery simply deepens as one gains more knowledge. You might also ask students at what point holding onto a romantic view of life becomes a refusal to grow up, to participate in adult life.

Using Specific Readings

"THE SANCTUARY OF SCHOOL," LYNDA BARRY

For Openers

Ask students to think and write about their havens. Is it a place (like school), an activity (like painting), a person (like Mrs. Lesane), a process (like slipping away in the dark)? Have these sanctuaries changed over the years? If so, how?

Collaborative Activity

Barry uses two main pairs of contrasting images: soundlessness versus noise, and darkness versus light. Divide the essay into five parts, and have each group take one part and analyze the dominant images in that section. As a class, discuss how the images above shift throughout the piece. For the last section of the essay, examine how Barry works the dominant images in with her discussion of the current educational funding crisis. Note how her argument grows from a childhood memory to a discussion of public policy.

Writer's Options

Like the visual artist that she has become, Barry creates some vivid pictures with her writing. She frames the essay with two images of voicelessness: In paragraph 2, Barry and her brother watch television without sound, and by paragraph 24, children across the country are speaking together a pledge that Barry believes falls on deaf ears. In your journal, write down some images you can use for a similar effect. Go through each of the senses (sight, sound, touch, smell, taste) and create contrasts and paradoxes possible for your sensory experiences.

"GRADUATION," MAYA ANGELOU

For Openers

Ask students to think of a time when they were happy or proud and something happened to take away that feeling. How did they react? What steps did they take to regain their optimism or self-respect?

Teaching Strategies

1. In paragraph 2 Angelou compares the physical plant of Lafayette Training School with that of the neighboring white school. What are the differences, and what community priorities are implied by the contrasts in the physical surroundings of the two schools?

2. Ask students to identify examples of Angelou's use of figurative language throughout the story ("The man's dead words fell like bricks" [paragraph 42], for example). Ask them to discuss how these images affect their reactions to the story.

Collaborative Activity

Ask students working in pairs, to discuss how one of the other characters (Donleavy, Henry, the narrator's mother, for example) might have reacted to the events of the story. (Assign different groups to focus on different characters.) In class discussion, identify similarities and differences among the characters' reactions.

Writer's Options

Write a graduation address that you might deliver if you were the Stamps High School valedictorian the following year.

Multimedia Resources

Bring in an audio tape of "Lift Every Voice and Sing," and play it for the class. Compare it with the "Star-Spangled Banner." How are the melodies different? How are the images different? Are their purposes different?

Suggested Answers for Responding to Reading Questions

1. Some educators still have limited expectations for minority groups.

2. Donleavy's speech is really eye-opening: It shows the graduates the limits placed on them by the white culture. Angelou's optimism and hope for the future are shaken as she listens to him, but they return when Henry Reeds sings the Negro national anthem.

3. Angelou's statement means that the spirits of the students in the audience were not repressed or crushed. Despite the speech, an indomitable spirit, maintained by their community, could not be broken. Therefore, they would rise above racism and prejudices and be "on top."

Additional Questions for Responding to Reading

1. How does Angelou convey a sense of community in this essay?

2. Opportunities were very limited for African-Americans in Angelou's early years. Do you think this situation has changed?

3. The last few paragraphs serve as both a conclusion and a commentary for the essay. What message does Angelou convey in her conclusion?

"SAVAGE INEQUALITIES," JONATHAN KOZOL

For Openers

The principal of the affluent PS 24 in Riverdale understands and is concerned about the conditions at the other two schools in Kozol's study. In the final paragraph, the principal asks; "They enter school five years behind...what *do* they get?" What do those students get?

Teaching Strategy

At the end of Kozol's visit to PS 261, a teacher says, in reference to their dilapidated physical plant, "They don't comment on it but you see it in their eyes. They understand." Discuss this quotation with your class in the context of William A. Henry's "In Defense of Elitism," and then in the context of Maya Angelou's "Graduation."

Collaborative Activity

The physical proximity of the schools Kozol studies is important to him, as are the detailed descriptions he gives of each school. In groups, students can conduct a similar study on your campus. Which buildings, although near each other, have dramatically different qualities? Which ones are better equipped, and which buildings' facilities are in need of updating? Once they have done this "fieldwork," the entire class can discuss what message this sends about the funding priorities at your institution.

Writer's Options

In paragraph 10 Kozol locates the poverty-stricken P.S. 261 in relation to a mortician's office. Indeed, the image is supplied by the person giving him directions to the school. Clearly, the image frames his view of the school as well, as Kozol sees the mortician's office as a cruel and constant reminder of the plight of this particular institution. In your journal, brainstorm for images that define an educational setting you know well: What other building or service might serve as a metaphor for a class you took?

Multimedia Resources

Students outside of the New York City area might have a difficult time following Kozol's initial description of how the New York schools fit together. Bring in a large map for the class, and show how the island of Manhattan is surrounded by the other boroughs, which in turn are surrounded by the suburbs. Color-code the economic status of each section according to Kozol's description to vividly illustrate just how physically near to one another the three sections in question are.

Suggested Answers for Responding to Reading Questions

1. For another perspective on the American obligation to provide "the means to compete," see William A. Henry, III, "In Defense of Elitism."

2. By including other examples from other school districts, the situation Kozol writes about could not be interpreted as an

isolated situation, or that his view is idiosyncratic. Providing a wide range of evidence is more effectively persuasive.

3. Open for discussion.

Additional Questions for Responding to Reading

1. Kozol describes the school in the worst condition first; he ends by describing the most affluent. How does this order shape the way you read the chapter? Would the force of the piece be weakened if Kozol reversed the order?

2. The word *special* takes on different meanings in different settings; how does the definition of *special* shift within PS 24?

3. Why do you think the teachers at the poorest schools are considered "the worst"? Are they bad teachers?

"COLLEGE PRESSURES," WILLIAM ZINSSER

For Openers

Ask students to list the reasons why they are enrolled in college. Follow with a discussion of the many reasons people attend college

Teaching Strategy

Using your college's course catalog, discuss the range of course offerings. Ask students to consider what clues the catalog gives to the purpose of a college education. According to the catalog, what function does the college see itself as serving?

Collaborative Activity

Divide the class into four groups, one for each of the types of pressure Zinsser identifies. Then have each group identify the sources, manifestations, and coping strategies of one kind of pressure. Finally, ask groups to share their findings with the class.

Writer's Options

1. Write a response to Zinsser in which you discuss how relevant you feel his pressures are to your own experience in college.

2. Write a letter to your college president offering a suggestion that would improve student life at your school.

1. Answers will vary, but students should be able to cite at least one point of difference.

2. Students at Yale, a highly selective school, are perhaps not representative, but certainly most college students share some common goals and similar pressures to achieve those goals. Students in less elite institutions may face greater economic pressures or have to struggle harder to succeed academically.

3. Zinsser's view of the ideal college is one where students have the time and the inclination to find out what interests and excites them; they should be able to take a wide variety of courses to see what they like; they should be free to explore those extra-curricular activities that most intrigue them. True to his position at Yale, Zinsser believes in the ideal of the "liberal arts."

Additional Questions for Responding to Reading

1. If you were not in college, what would you be doing?

2. What are the advantages and disadvantages of attending college? Of alternative options.

"IN DEFENSE OF ELITISM," WILLIAM A. HENRY III

For Openers

Ask your students if they would be able to attend college or would have been admitted to college if Henry's plan were enacted. If not, what would they be doing?

Teaching Strategy

Henry makes a nice stylistic move in paragraph 11, when he takes a commonly agreed-upon point and rewrites it to question its value: "As a society we consider it cruel not to give them every chance at success. It might be more cruel to let them go on fooling themselves." Use this as an opportunity to teach about appealing to the common ground between the writer and the audience.

In paragraph 6 Henry asks, "Why do people go to college?" Answer this question for yourself.

Multimedia Resources

If you did not discuss the film *Stand and Deliver* at the beginning of this chapter, now might be a good time to introduce it to your students. In this film, the math teacher refuses to let the level of his teaching sink to the level of expectations the administration has for the students. At the same time, however, he sees college as a way out for his students. Where might that teacher agree with Henry, and where might he disagree?

Suggested Answers for Responding to Reading Questions

1. Both. In using the term elitism, Henry understands its snob appeal but is also trying to give the word some positive connotations in a society devoted to equality.

2. Answers may vary.

3. Henry believes that the best will rise to the top. He fears that too much has been done to guarantee success to all, rather than to guarantee equal access to that success for those willing to spend the time, money and effort. Agreement will vary.

Additional Questions for Responding to Reading

1. Do you agree with Henry's assertion that "many students vociferously object to being marked down for spelling or grammar"?

2. What effect does Henry's word choice have on you as a reader? For example, he describes those who are not elite as "mediocrities," "idle," and "untalented." Do you agree with those descriptions, or would you choose other labels? If so, how would you describe the nonelite?

"SEPARATE IS BETTER," SUSAN ESTRICH

For Openers

Estrich's title, "Separate is Better," implies that separate is always better. Discuss whether her essay supports this assumption. If not, decide upon a more fitting title for the essay.

Collaborative Activity

The phrase in paragraph 8, "women could do anything, because no one told us we couldn't," sounds like an excerpt from the catalog descriptions of several women's colleges. In groups, ask students to study your own university's catalog and brochures and list the positive descriptions used to "advertise" your school. Students can paraphrase or rewrite these catch phrases to present reality as they see it. After they have been rewritten, share the new phrases with the class.

Writer's Options

Estrich uses her baton twirling as a symbol of how much she has changed. Is there an activity in your life that you can build into an icon of days you have left behind? Try to use this image in your next paper.

Teaching Strategy

At the end of paragraph 3, Estrich claims that the real problem wasn't with "the PSAT, but me, and my school." By refusing to blame anybody but herself for her inability to receive a Merit Scholarship, Estrich is accepting certain ideas about the responsibility of individuals in the educational system. What are those assumptions? Can you come up with some clichés about individual responsibility that Estrich would agree with?

Suggested Answers for Responding to Reading Questions

1. Unfortunately, Estrich does not fully explain how a single-sex high school would help African-American boys. She does, however, build an argument that can be extended to other groups besides women.

2. This is an example of soft evidence and hard facts. Each is persuasive since the personal experiences humanize the issue, and statistics allow one to objectively analyze the problem.

3. Answers may vary.

Additional Questions for Responding to Reading

1. Estrich argues that public policy decisions unwittingly make the problem worse, by demanding that government-funded institutions guarantee equal access, regardless of gender, color, and so on. Do you think the benefits outweigh the shortcomings of this legislation, or do you agree with Estrich's point? Explain your answer.

2. In her final paragraph, Estrich asks that educators be trusted to abandon programs that do not work. Do you agree that these kinds of decisions can be made by educators? Why or why not?

"THE VISIGOTHS IN TWEED," DINESH D'SOUZA

For Openers

D'Souza ends his essay with a call to action, a plea to interested people to stay informed about issues at the institutions they respect. This advice is as important for those who agree with D'Souza as it is for those who disagree with him. In class, discuss some of the issues currently under consideration at your university.

Teaching Strategy

Explain that the term *Visigoth* (in the title of the essay) refers to a member of the westernmost branch of the Goths, the ancient Germanic monarchy that invaded parts of the Roman Empire. A secondary definition for Goth is a rude, barbaric person. Why does D'Souza use this term in the title—and why doesn't he refer to it within the essay?

Collaborative Activity

D'Souza makes several claims about the changing core curriculum at colleges and universities throughout the country. Compare his characterization of your own experience. What are the core requirements at your institution? What courses are required at schools your friends attend? Bring together this information as a class, and check to see if D'Souza's descriptions seem accurate.

Writer's Options

Many of the terms D'Souza uses to describe the academic left are emotionally loaded. Find a paragraph that contains some of the most judgmental terms and rewrite it so it sounds more objective and less inflammatory. Which version do you prefer and why?

Multimedia Resources

D'Souza criticizes a Duke University professor for using the movie *The Godfather* as a metaphor for the state of American business. Bring in scenes from *The Godfather* and scenes from either *Wall Street, Disclosure,* or *Working Girl.* Does D'Souza's criticism seem like a valid argument to make, or does it detract from some of the other points he makes throughout the essay?

Suggested Answers for Responding to Reading Questions

1. D'Souza is referring to the changes in university and college curricula as a revolution. He is of the opinion that a widespread interest in issues of multiculturalism and diversity is causing an upheaval in American higher education. He also indirectly suggests that this is a political move on the part of the faculty.

2. Answers will vary.

3. D'Souza suggests that white males are forced to "confess" their guilt before they will be accepted by their professors. The over-dramatic comparison to Stalin seeks to startle the reader into a position that would protest the plight of today's students.

Additional Questions for Responding to Reading

1. How does your knowledge of D'Souza's ethnicity change the way you read this essay? Would it make a difference if he were born in the United States — or in the former Soviet Union — instead of in India?

2. Compare D'Souza's argument with Henry's "In Defense of Elitism." The two writers identify a similar problem, but they assign blame to different groups. Which writer do you agree

with, if you agree with either? Explain their positions, and your responses to their positions.

Focus: What is Good Teaching?

All institutions are redefining themselves in order to be ready for the next century. Ask students to consider what is essential in the American institution of higher education. As multimedia, distance learning, and computer communication become more familiar to both student and faculty, what is central to learning and engagement with course work?

"VIRTUAL STUDENTS, DIGITAL CLASSROOM," NEIL POSTMAN

For Openers

1. Ask students to consider how becoming educated can be both passive and active on their part.
2. Ask students to consider whether they come to college to gather information or to acquire the ability and discipline of interpreting and organizing information.

Collaborative Activity

Ask students to discuss in groups how they expected college classes to be different from high school classes when they were still in high school. Have those expectations been met?

Teaching Strategies

1. Discuss the effect of numerical statistics in paragraphs 9 and 10. Are these large numbers impressive and convincing? Ask students to consider other arguments or situations where statistics are a necessary part of their decision to accept or not accept an opinion.
2. Postman creates an analogy between driver education and technological education. Discuss the value of an extended analogy. Question the effectiveness of this analogy. You might ask students to highlight all references to driver education to see how it is a repeated element that gives unity to the essay. Ask students to consider other analogies.

Writer's Options

Choose a topic that is of interest to you and your classmates. Using only a computer, research that topic. Does the research result in an overwhelming amount of information? Do you agree that a computer-based researcher will tend to gather information rather than question it?

Multimedia Resources

Find the Prentice Hall Web Page. Explore different teaching supports found on different web sites. How do these computer-based teaching supports conflict or support Postman's assertions?

Suggested Answers for Responding to Reading Questions

1. Postman thinks technology can enhance an education but it cannot replace it. Although computer technology can bring information to children, it does not necessarily teach how to process information, nor does its speed support reflective though and writing.

2. Postman believes that schools should teach students more than how to operate a computer. Postman holds that using a computer means understanding the cultural effects of computer-based education. Education should be based on inquiry; students would not merely gather information, but question the information gathered. Postman warns that opportunities to work in groups must not be sacrificed to individual work at computer stations. Interaction with other students is more important than interaction with a computer screen. Postman also warns that computer technology may not be accessible to some economic groups. Technology does not solve all educational problems.

3. Technology cannot solve all social problems. Education cannot hope that any technology will solve long-standing problems. Students must resist merely operating equipment, but rather students must use technology with discretion, and they must question the technology itself. This view is neither pessimistic or optimistic.

1. How has your education with computers been different from those who have been educated without computers? For example, compare your learning situation to that of your parents.
2. Why does Postman include the poem in paragraph 17?

"COMMONPLACES ABOUT TEACHING: SECOND THOUGHTS," BOBBY FONG

For Openers

Since this article was written for teachers, explain the term "pedagogy" and describe what academic conferences are like.

Teaching Strategies

1. Consider commonplaces that students may have about the university. For example, "If you make the teacher cry, you'll get an A" or "Show progress. Don't do your best work right away; this way the teachers feel good about themselves." List the commonplaces on the board.
2. To review the essay for "hard facts" and "soft stories" ask students to highlight examples of each with two different colors. What is the predominating persuasive kind of evidence in this essay?

Collaborative Activity

Divide class into six groups, assigning a commonplace to each group. Have each group discuss what surprised them in that part of the essay. To what extent, based on their shared experiences, does the group agree or disagree with Fong?

Writer's Options

1. Fong says that a lecture is like a guided walk in the forest. Choose a different metaphor and explain the virtues and failings of lectures by means of that metaphor. Take into consideration not only the lecturer but also the students and the setting.
2. Using the class discussion on university culture as a springboard, write an essay that explains your opinion and

description of a university culture. Use the different essays and your own experience for support.

3. Write an essay that states your own goals for your college career. In what ways will you resist becoming a silent voice in your own education?

Multimedia Resources

Ask different groups of students to view *Stand by Me, Dangerous Minds,* and *Renaissance Man.* Ask students to pay particular attention to student-teacher relationships and how the teacher in each movie makes a difference in what is learned. How is language an important part of the students' educational experience?

Suggested Answers for Responding to Reading Questions

1. Commonplaces are statements that are assumed to be true and are stated without challenge. Perhaps commonplaces seemed true to enough people that it became an accepted generalization.

2. Fong is claiming that college provides an opportunity for students to be exposed to new ideas and experiences that are not part of their at-home culture and to reexamine that home culture. Students may share their own reflections.

3. "Nowhere do you find more enthusiasm for the god of technology than among education..." or "...university culture is ever more devoted to consumption and entertainment...."

Additional Questions for Responding to Reading

1. What is the difference between the words reflexive and reflective?

2. How do you know Fong is an experienced teacher? How does he establish his credibility?

"ON THE USES OF A LIBERAL EDUCATION," MARK EDMUNDSON

For Openers

Ask students to list 20 words and phrases that describe the attributes of excellent teaching.

Teaching Strategies

Ask students to number a page from 1 to 59 (for each paragraph in the essay), leaving room to write a sentence or two next to each number. Ask students to write a summary phrase or sentence about that paragraph. Ask students to notice when Edmundson uses questions in his essay.

Collaborative Activities

1. Ask students to observe a part of campus, with each group member choosing a different time of the day from other group members. One group may observe the library computer area, another the dining hall, a department lobby, a parking deck, a main walkway, the student center, or the bookstore. For about twenty minutes, write field notes of observations. Ask the members of the group to compare notes and compile a composite report for the rest of the class. After each group presents its findings, discuss whether the campus is "laid back," apathetic, and "cool."

2. Ask students to videotape a class, with permission of the class and instructor. Then ask them to analyze what they see in terms of Edmundson's essay.

Writer's Options

1. Describe the qualities of an ideal teacher-student relationship. Considering your description, what does this relationship assume about the university community? Finally, how does your view relate to Edmundson's view?

2. Research student organizations on your campus. What are their purposes? Consider intramural sports, environmental groups, and political organizations.

Multimedia Resources

1. Choose a variety of types of universities from private to public schools. Review their web pages. What seems to be the lure or "sales angle"? Do the web pages support Edmunson's claim? Is rigor or entertainment the larger part of the web pages? Are Edmunson's descriptions in paragraph 26 accurate?

2. The comic strip "Doonsbury" often satirizes the very condition Edmundson describes. Bring published collections of this comic strip to class for analysis of the humor.

Suggested Answers for Responding to Reading Questions

1. Education becomes a matter of cost and return. A student's return is a grade, not necessarily a consideration of ideas. One buys into a course and earns a grade. The media promotes a passive engagement with cynicism as the norm in comedy and commentary.

2. Edmundson holds that recreational areas such as sports centers, physical fitness centers, and gardens are dominating the campus environment. Edmundson sees this as a need to provide entertainment and a sense of leisure to the students in order to make the school more marketable to the widest possible range of students.

3. Edmundson thinks that instructors must create courses and presentations that challenge rather than entertain their students. Students must seek a deeper meaning and connection to courses, not accept professors' unchallenged presentations. Students may want to consider the role of student groups on campus.

Additional Questions for Responding to Reading

1. Edmundson says that faculty should "defy student convictions and affront them occasionally." Find places where Edmundson is offensive in his description of students. What is your response to his description?

2. Are only students guilty of a consumer approach to education or are faculty also involved?

3. Edmundson speaks of an "acculturation" into academia. As a college student how do you see yourself being drawn into your campus culture?

4. Were the '60s really very different from today's campus environment?

CHAPTER 3

THE POLITICS OF LANGUAGE

Setting Up the Unit: Using Language Play

for Rhetorical Variety

The selections in this chapter offer students many avenues for exploring the roles that language plays in their lives. Some essays address stylistic issues directly, focusing on word choice, literacy, bilingualism, and the differences between public languages and private ones. All of them discuss, directly or indirectly, the ways that language helps shape our social interactions, including the ways in which we are characterized by others, or the ways in which we characterize them, based on language cues. It is this social character of language that makes it political, not only on a national or world scale, but a personal scale as well.

Since language is the subject matter as well as the medium of expression, it follows that many of the writers whose pieces are included in this chapter are self-conscious about the ways in which they are using language — more so than in other chapters of the book. As your students read these selections, they will be exposed to a wide range of experimental uses of language, different from those dictated by Standard Written English. And students will be able to judge for themselves whether or not these techniques are successful.

Stepping away from the political implications of language choice for a moment, you may find it useful to talk with students about the types of language experimentation that might be permissible in your class. Departures from the norm have two main effects: They show the student's individuality, and they allow for more "creative" types of writing to infuse expository compositions. Students may want to learn more about both of these strategies. This chapter offers you, as the instructor, the chance to define the language boundaries in your classroom; at the same time, it offers students a chance to think about the flexibility they might have within those boundaries.

For example, some of the writers included in this chapter sprinkle in words or phrases from other languages, often their native tongues. These phrases add a flavor to the writing that makes readers notice and slow down. They mark the writer as somebody with something interesting and different to say. Likewise, the use of dialogue within an essay can break the flow of a description, making readers pause for a moment. Other essays in this chapter show various meanings of a single work; Malcolm X, for instance, spins out numerous possible meanings of specific terms.

Students can experiment stylistically with images and metaphors and with dialects and levels of diction. Writing assignments growing out of this chapter's readings can build on the language play evident in many of them. Either in journals or in full-length papers, students should be willing to try something new. They can interview others and quote their words. They can mix voices, clarifying their own distinctions between public and private languages. Those who speak other languages and dialects can try incorporating phrases into their writing. And any student, at any time, can examine individual words and focus on their rich connotations and changing relationships to other words. In short, this chapter provides an opportunity to experiment with and perhaps depart from formal language conventions.

Confronting the Issues

Option 1: Constructing Contexts

Many students do not understand that words have the power to convey attitudes and value judgments through connotative and denotative meanings. For example, describing an executive as *pushy* implies that he or she is inappropriately aggressive. Describing that same executive as *assertive* or *forceful* implies that his or her personality traits are praiseworthy.

To dispel the mistaken impression that words with similar meanings are interchangeable, do the following in-class exercise. Duplicate a page or two from either *Webster's* or *Roget's Thesaurus* and ask students to consider the synonyms that are listed. For example, the following synonyms are listed in one thesaurus under the entry *crisis: juncture, contingency, crossroads, emergency, exigency, pass, pinch, strait, turning point, and zero hour.* After some discussion it should become clear that these words all have differences in denotation and connotation. Adding each of them to the following sentence should further illustrate this point: *Declining enrollment and decreasing tuition revenue have caused a financial _____ in many colleges.* Clearly, most of these "synonyms" do not convey exactly the same meaning as *crisis.*

You can go on to point out that writers are especially aware of the subtle differences that exist between words. The poet Louis Zukofsky called language the "finer mathematics" and writer Dorothy Parker called wisecracking "calisthenics with words." Once students are sensitized to the importance writers place on words, they can begin to understand the struggle that goes on in every selection in this book to find the one word that has just the right shade of meaning to convey the writer's ideas and feelings.

Option 2: Community Involvement

Most of the student-writing samples that begin this chapter refer to specific word choice in the public arena, particularly those word choices considered inflammatory (at worst) or insensitive (at best). For this project, then, students can bring in examples of writing from newspapers or magazine that they would like to rewrite in a more inclusive or sensitive way. These revisions could hinge on a single word choice, such as the difference between "sexual preference" and "sexual orientation" in a public statement ensuring equal rights, or news reports about Greg Louganis saying he "admitted" he has AIDS—as opposed to "disclosed," or "said."

Option 3: Cultural Critique

Bring in a recent episode of *Sesame Street*, preferably one in which characters discuss language, spelling, or the alphabet. As a class, analyze how the subject matter is presented: What messages are being sent about race, class, gender, sexual orientation, religion, age, and education? List these messages out on the blackboard, and discuss the implications that these "coded messages" have on language use and language acquisition. You might want to tell your students that at least one country rejected the program, saying *Sesame Street* attempted to "brainwash" children.

The video series *The Story of English* devotes an entire episode to sociolinguistics and the issues discussed throughout this chapter. Set up class discussions by watching part—or all—of this episode. If you are in a class where students are likely to find the writing in this chapter too "nonacademic," or perhaps too politicized, this video will help establish that this inquiry into the political nature of language has a long history and is an integral part of English studies.

Teaching "Two Perspectives"

Before they read Douglass and Malcolm X, ask students to look at the student voices that open this chapter. Ask them to pick one of the students, and write a response. After reading the next two essays, students might want to go back to revise what they wrote.

"LEARNING TO READ AND WRITE," FREDERICK DOUGLASS

For Openers

Douglass identifies books and documents that he encountered while trying to read and write. Ask students to discuss what books have most influenced them. In what ways was their perception of life altered by their reading?

Teaching Strategies

1. Douglass writes with sarcasm in paragraph 4, "...it is almost an unpardonable offense to teach slaves to read in this Christian country." What prompts his ironic tone? How is the refusal to teach reading and writing to slaves inconsistent with the ideals of Christianity?

2. Discuss the background of slavery with students. Point out to them that slaveholders did not consider slaves human beings. In addition, let students know that slaves were thought to be incapable of learning or of embracing moral values. (To grant them these attributes would be to accept their humanness and thereby undercut the rationale for keeping them in bondage.) Throughout his life as a free man these attitudes haunted Douglass. Many contemporaries considered him a freak of nature, anomalous to what they knew to be the actual nature of enslaved African-Americans.

50

Collaborative Activity

Divide students into groups and ask them to retell Douglass's story of how he learned to read using one of the following perspectives: Mrs. Hugh, Mr. Hugh, Thomas Hugh, a fellow slave on the Hugh plantation, a freed slave after the Civil War, and Douglass himself in one of his later autobiographies. Share the new versions with the class.

Writer's Options

1. Discuss the books that have changed the way you look at the world and yourself.

2. Examine how your life would be different if you could not read or write.

3. Both Frederick Douglass and Richard Wright ("The Library Card," page 26) used reading as a means of overcoming restrictions placed upon them by society. Write about a means you used to overcome restrictions placed on you by society.

Multimedia Resources

In a scene from the recent movie release of Mary Shelley's *Frankenstein*, Robert DeNiro as the "monster" learns to read and write in much the same way Douglass describes in paragraph 8. Show the scene to your students, and explain that Shelley was writing at approximately the same time Douglass was growing up. Her book provides a social commentary about the dangers of over-industrialization at the expense of humanity, through institutions like slavery. Shelley never mentions slavery; still, how does the way in which both learners succeed in reading comment on the plight of the slaves? And how do their similar methods for obtaining a secret education comment on the rules that their society imposed?

"A HOMEMADE EDUCATION," MALCOLM X

For Openers

In the final paragraph, Malcolm X describes the advantages of learning in a prison rather than in a college. What do your students think of this distinction?

Teaching Strategies

1. Discuss the problems caused by Malcolm X's inability to express himself in a language other than slang. Do style and level of diction really matter as long as one is able to get a message across?

2. Poll students on the following questions: How often do you use a dictionary? For what purposes? Has reading this essay affected the way you view the dictionary?

3. Ask students how much African-American history was presented to them in the course of their education. Do African-Americans have reason to object to the way American history is taught? Do other groups also have reason to complain?

4. Discuss the subjective nature of history. Make clear that human beings record it and that the details preserved depend entirely on what the person recording it perceives as important.

5. Discuss the following with your students:

   Paragraph 24: What conclusion is made "clear"? What is Malcolm X implying?

   Paragraph 27, line 9: To what or to whom is Malcolm X referring?

   Paragraph 38: Have students heard the phrase "new world order" before? What does Malcolm X mean when he uses it?

   Paragraph 42-43: Ask students whether they agree with the reasoning of this digression.

Collaborative Activity

In recent years, some parents have tried to remove their children from the public schools and educate them at home. Stage a debate on the superiority of this kind of "homemade education" over an institutionalized one. Students should support their arguments with evidence from Malcolm X's essay as well as from their personal experience.

Writer's Options

1. Using the information you generated during your collaborative learning activity, argue for or against the superiority of a "homemade education."

2. In addition to learning facts in college, you also learn a number of specialized vocabularies. Write a description of the

specialized vocabularies you have learned in two or three of your classes. In what way have these vocabularies enabled you to express new ideas?

Multimedia Resources

Spike Lee's film version of *Malcolm X*, based on the book from which this excerpt was taken, presents a slightly different picture of the role dictionaries played in Malcolm X's prison education. In class, show the scenes leading up to Malcolm X's trip to the prison library with his new-found friend, where they discuss the "prison of the mind" that envelops African-Americans both inside and outside of jail. Continue viewing through the dictionary scene, in which Malcolm X discovers for himself the "white man's" definitions of the words *black* and *white*. He then comes to see how loaded other definitions are. Afterwards, discuss what is different about the print and film versions, what effect those differences have, and possible reasons for the choices Lee made.

Two Perspectives: Suggested Answers for Responding to Reading Questions

1.   Douglass's ideas about slavery were formed by his actual experiences with it, but learning to read and write made him aware of the possibilities of achieving freedom, both physically and intellectually. Although Douglass knew slavery was wrong even before he learned to read and write, these skills focused his ideas so he understood why slavery was wrong. Language empowered him by enabling him to become a powerful spokesperson for the abolitionist movement.

2.   Malcolm X wanted to increase his skills so that he could express his ideas in writing. Not only did he learn vocabulary and writing skills, but he also learned about bigotry and how history had been "whitened." He also learned more about the horrors of slavery.

3.   Both Douglass and Malcolm X were self-taught. However, Douglass was discouraged by his slaveholders from learning how to read and write while Malcolm X was encouraged to read and write as a sign of his interest in rehabilitating himself.

4.   Answers will vary. Because it makes a person aware of himself or herself as an individual with certain inherent rights, education is inimical to slavery. Slaveholders feared that slaves would organize and revolt if they were educated.

Prison officials did not encourage Malcolm X because he would stay up all night reading instead of following the lights-out rule. The would prefer compliance to their rules over his education.

5. Disadvantages to self-taught methods include errors being undetected and not corrected and no outside encouragement. Douglass missed having teachers to correct his work. Malcolm X celebrated that he did not have distractions of outsiders so he could focus on his studies.

6. Both essays decry the sub-human conditions in which those of the Black race were placed. Both describe discrimination and prejudice. These topics continue to be worthy of discussion.

Using Specific Readings

"FROM OUTSIDE, IN," BARBARA MELLIX

For Openers

Why does Mellix begin and end this essay with connections to her family? Does it matter that at the beginning she speaks as a mother, and at the end she speaks as a daughter?

Teaching Strategy

If your students have a difficult time relating to the issue of dialect here, have them write down expressions as they would say them to their friends. They, too, have different languages they use in different situations, depending on their audience and their purpose. Telling a story to a best friend is different from telling it to the dean of the college, so students should be able to understand the distinctions that Mellix draws.

Collaborative Activity

In paragraph 11 Mellix shows the standard English sentences she might have written as a child, and she translates them into Black English. Divide the class into five groups, and ask each group to analyze the two versions of one sentence. How do the translations differ? In some cases, the meaning is slightly altered, so why might the standard English sentence be more "correct"?

Writer's Options

In paragraph 7 Mellix uses an unusual metaphor "I had taken out my English and put it on as I did my church clothes, and I felt as if I were wearing my Sunday best in the middle of he week." Relate the topic of your next paper to a metaphor, experimenting with at least three different ones. How does each change the nature of what you are saying?

Multimedia Resources

The 1980 comedy *Airplane!* has a scene that provides subtitles for two men speaking Black English. A nice old little lady who says she speaks "Jive" offers to translate his words into standard English for the flight attendant. Showing this scene might be an interesting and novel way to introduce and reinforce the issues Mellix raises.

Suggested Answers for Responding to Reading Questions

1. Mellix shows in her essay how the language one speaks can determine how one is regarded by others in regard to class. She also shows how black English is celebrated in one community and avoided in another.

2. Mellix views standard English as the language of power, of the ruling institutions. She associates Black English with a family language, best kept in the confines of the home.

3. Since Standard English is considered as the language of power by Mellix, then those who can speak it share powerful positions in society. However, she also regards a loss of herself and her heritage.

Additional Questions for Responding to Reading

1. How many different languages do you speak? Make lists of words and terms you use only in certain groups.

2. What words do you see in magazines or hear in lyrics that are borrowed from non-standard English? Do you think these words will eventually be accepted by the mainstream and adopted?

For Openers

Tan reacts against the term *limited English* in paragraph 8, saying that it seems as if everything is limited, including people's perceptions of the limited English speaker." Do your students agree with this perception? Has someone they know been treated as "limited" because he or she does not speak "proper" English?

Teaching Strategy

In paragraph 14 Tan says of her mother, "She said she had spoken very good English, her best English, no mistakes." Discuss Tan's purpose for using this type of indirect quotation. Do our students think strategy is effective? Encourage them to look for additional examples of indirect quotations in the essay.

Writer's Options

At the end of her essay, Tan says that she writes most of her fiction with her mother in mind. Do you have a reader in your head who guides you and keeps you honest? If so, who is it? If not, who might it be? Write a description of this person (or persona), and explain how keeping that person in mind changes the way you write.

Suggested Answers for Responding to Reading Questions

1. Answers will vary, but ask students to consider how sincerity is a way to gain credibility and also help build trust with the audience.

2. Answers will vary.

3. Since the SAT is a timed exam, the faster your recall of certain words, phrases, and concepts, the better your chances of getting more correct answers.

Additional Questions for Responding to Reading

1. In paragraph 3 Tan describes how she feels her writing is burdened with sophisticated grammatical forms. What does she mean? Is it?

2. Tan writes that she hopes to capture the rhythms of her mother's speech in her writing. Reread the essay, and decide whether or

not she succeeds. If she does not, where and how might you rewrite it?

"THE HUMAN COST OF AN ILLITERATE SOCIETY," JONATHAN KOZOL

For Openers

Kozol does not talk about the causes of illiteracy. In class, discuss the reasons illiteracy is so widespread. What do your students identify as the causes? What do they think we can do about it?

Teaching Strategy

One reason students graduate from high school without knowing how to read is learning disabilities that go undiagnosed; another reason is insufficient means to cope with a disability that has been diagnosed. Use this opportunity to discuss different learning disabilities with your students. What accommodations, if any, do they believe should be made for students with such disabilities?

Writer's Options

In paragraph 6 Kozol describes a recurring nightmare that dramatizes what he feels illiteracy must be like. Try this yourself: Imagine yourself in a situation where you are unable to communicate for some reason. What specific obstacles might you face?

Multimedia Resources

A scene from the 1989 film *Stanley & Iris* shows the frustrations of an illiterate man who is learning to read for the first time as an adult. To introduce a short in-class discussion, show the scene to reinforce and supplement the images Kozol presents. For a longer activity, have a screening of the entire film and contrast its message with Kozol's. In the movie, Stanley (played by Robert DeNiro) becomes wealthy and successful after Iris (played by Jane Fonda) teaches him to read, perpetuating the ideal that education is a guarantee of financial well-being and that learning to read will solve all or most of society's problems. Ask students if they think that these assumptions are reasonable.

1. Answers will vary.
2. Answers will vary.
3. Answers will vary.

Additional Questions for Responding to Reading

1. Kozol begins the essay by presenting one example after another of what illiterates cannot do. Afterwards, we read the actual words of the people who cannot read. What purpose does this organization serve? Would you read the essay differently if he integrated the voices with his examples? If he presented them first?

2. What is the effect of the Kozol's technique of repeating "Illiterates cannot..." "Illiterates cannot..." "Illiterates cannot..." in paragraphs 8 through 24? Does it dehumanize these people? Does it make their situation more compelling? Would you ever use this technique in your own writing?

"SEXISM IN ENGLISH: A 1990S UPDATE," ALLEEN PACE NILSEN

For Openers

What does the language used to describe men and women indicate about their respective roles in American Society?

Teaching Strategy

Discuss the following with your students:

Paragraph 6: Nilsen opens her argument by stating her position and supplying support. Of what value is such a strategy? Doesn't she risk offending her audience? What effect does her opening have on you?

Paragraphs 10-11: Account for Nilsen's findings in her exploration of eponyms and geographical names.

Paragraphs 17-18: What is your reaction to the changes Nilsen describes as having occurred since her 1970 study?

Paragraph 21: Today some women who marry keep their names. Do you think this behavior is appropriate? Why or why not? What would you think of a man willing to take his wife's name?

Paragraph 29: Why are behaviors that are perceived as "feminine" discouraged in young boys?

Collaborative Activity

Ask members of student groups to recall their upbringing and socialization as children with reference to sex roles. Ask them to list behaviors that were encouraged and those that were explicitly or implicitly discouraged. Direct them to look especially for subtleties (room decor, clothing, toys). Groups should present their findings and conclusions to the class.

Writer's Options

Write an essay about a time when you were discouraged — by parents, teachers, friends, or co-workers — from behaving in a particular way because such behavior was considered not to be consistent with your gender role. Include, if possible, the language used to discourage you.

Multimedia Resources

Now is another good time to look at *Sesame Street, Barney and Friends*, or any other appropriate children's television shows to see if their language use is consistent with what Nilsen describes. Contrast these shows with Saturday morning cartoons, and ask students to describe any contradictions they see there.

Suggested Answers for Responding to Reading Questions

1. Nilsen believes that a culture's language provides evidence of the thoughts that exist in people's minds. As Nilsen says, language is like "an X-ray in providing visible evidence of invisible thoughts."

2. Yes, she provides enough examples. The examples demonstrate that basic assumptions, however unrealistic, continue to be passed on from generation to generation.

3. Yes, but with great difficulty. Feminists must change not only the language of a culture, but also the underlying attitudes that are reflected by that language.

"Politics and the English Language," George Orwell

For Openers

1. Discuss recently coined political phrases, such as *new world order, a thousand points of light, politically correct,* and *litmus test.* Point out the ambiguity of these phrases and question their purposes.

2. Ask students to supply connotations and denotations for the following labels: *gay, liberal, conservative,* and *feminist.*

Teaching Strategies

1. "Politics and the English Language" was written long before the current pedagogical "writing as a process" theories were formulated or respected. Orwell's advice to students reflects certain approaches to writing that we call "Current-Traditional Rhetoric." One of the perils of following Orwell's advice too closely is the phenomenon of premature editing, which can often exacerbate a pre-existing writer's block. Stress to your students places within the writing process where Orwell's advice makes most sense. Keep in mind, too, that in writing pedagogy today, many of our values and assumptions as teachers of writing have changed in the fifty years since this piece was written. Remember, too, that Orwell freely admits to breaking his own "rules" to suit certain purposes, and this can be an important point to reinforce with students.

2. Discuss the following with your students:

   Paragraph 2: What cause-and-effect relationship does Orwell discuss here? Explain the reasoning behind his analysis and decide whether you agree.

   Paragraph 4: To what "certain topics" does Orwell refer?

   Paragraph 5: An interesting phenomenon some composition teachers have lately observed is the distortion of worn-out phrases by student writers who have heard the phrases widely used but have never seen them in print: for example, *take for granite* instead of *take for granted* and *It's a doggie dog* world instead of *It's a dog-eat-dog world.* Note the difference in meanings between the phrases. Are such distortions cause for amusement, or do they have serious implications?

Paragraph 6: This is a good opportunity to discuss the problems of the passive voice. For example, what is the difference between the following two phrases: *Taxes must be raised* and *We must raise taxes*? The active-voice phrase has a human subject. The passive -voice phrase does not; therefore, it allows the speaker to avoid responsibility for an action or for what is being said.

Paragraph 7: Emphasize that here Orwell attacks pretentious diction, not the varied, creative vocabulary necessary for effective communication.

Paragraph 8: Return to the For Openers discussion; reintroduce the terms and apply them in the context of this passage.

Paragraph 10: Why do you think Orwell chose a verse from Ecclesiastics for his parody? Could another kind of quotation have served as well?

Collaborative Activity

Before discussing the essay, divide students into five groups. Assign one of Orwell's sample passages to each group, and ask groups to judge the quality of the samples. After discussing the essay, reassemble the groups and direct them to revise their passages.

Writer's Options

Write a translation of a Bible passage or a well-known fable or fairy tale in a style similar to Orwell's translation in paragraph 10.

Multimedia Resources

Discuss the abuses of language to which Orwell was reacting in this essay by bringing in examples and letting students analyze them. For instance, Nazi posters juxtaposed terms and pictures to promote hatred of the Jews; the former Soviet Union also manipulated language to maintain social control. Add American propaganda to these discussions as well, particularly recruitment posters from World War II, or those depicting Japanese in a frightening—even inhumane—light. Do your students see any difference between old propaganda posters and today's advertisements?

1. Answers will vary.
2. Politicians and governments use vague, inexact language to justify, but not to explain, their actions.
3. Answers will vary.

Additional Questions for Responding to Reading

1. Why did Orwell title his essay "Politics and the English Language"? What connections does he draw between the two?
2. Do you think Orwell is inventing a problem, or are his complaints justifiable? Explain your answer.

Focus: Should All Americans Speak English?

The following essays do not offer absolute answers, but show the variety of conditions in language acquisition. Ask students to reflect on their own experiences with learning another language. Discuss what *American* means by asking students to write a list of words and phrases they associate with that word. Discuss the derivations of the American language. Ask the class to describe variations on the language. Read over the student voices at the beginning of the chapter. What additional perspectives do those voices bring to the discussion?

"AGAINST A CONFUSION OF TONGUES," WILLIAM A. HENRY III

For Openers

Loaded words, words that carry strong connotations, can be subtly persuasive. Consider how the title suggests the biblical story of the Tower of Babel. Consider how such a connotation predisposes the reader to think that Henry is offering protection from such a situation. Bring in different magazines. Scan titles looking for literary allusions and loaded words.

Teaching Strategy

Create the following scenario, and discuss its relevance to Henry's essay:

Next semester all your courses will use computers. All assignments and handouts will be located on a home page for the course. You will turn in assignments on-line. Furthermore, you will be expected to use word-processing programs, data bases, spread sheets, multimedia presentations, on-line research, MUDS and MOOS. You will be expected to make a Web page for your undergraduate career. This computer literacy will be used, but not taught. You will learn by assimilating computer literacy from those around you who are computer literate. Your grade will be based on what you learn about the course material and readings as shown through the projects generated on the computer.

Collaborative Activity

Identify divergent backgrounds and cultures in your group. How is language part of individual cultures? Discuss the disadvantages and advantages of being able to simultaneously be part of the two cultures.

Writer's Options

1. Write a parody of Henry's essay using computer literacy as the replacement for English. Describe the problem of becoming "mainstream computer literate" while still going to school with paper and pencil.

2. Rodriguez thinks that "the discomfort of giving up the language of home is far less significant than the isolation of being unable to speak the language of the larger world"(11). Consider a "larger world" that you joined by learning the language of that group. Argue the value of what is lost and what is gained.

Multimedia Resources

The movie version of Amy Tan's novel *The Joy Luck Club* (1993) depicts the struggle to assimilate into the American mainstream while maintaining a connection to family tradition and culture. Watch this movie and discuss how language is part of the transition.

1. The image of the melting pot suggests that diversity disappears into a new identity or character. In this case, all the different ethnic groups melt together to create the American culture. Any image that is made of distinct pieces creating a new pattern will work to show how the American scene is a composite of distinct ethnic groups. The salad bowl offered in the essay is one popular image. Various stones or bricks in a wall or construction may help illustrate the point. Molecules are made from different atoms combining by sharing parts of themselves, but not all. A mosaic is another possible image.

2. Answers will vary. Point out that "to overstate" means exaggerating and using strong language. By this definition, Henry does not seem to be overstating his position.

3. Henry's essay views long-range effects of bilingual education. He not only talks about its impact in the schools, but also in the society and cities themselves. He relies extensively on expert testimony and to a much lesser extent on statistics. Amselle, on the other hand, is mostly involved with the children. He relies heavily on studies and statistics to prove his point that as an educational policy, bilingual programs are flawed.

Additional Questions for Responding to Reading

1. Henry lists "four ways for schools to teach students." How are the following paragraphs organized to study the strengths and weaknesses of each method?

2. The word *political* is used often throughout this essay. What connotations does this word have for you? Does Henry depend on those connotations for his argument, or is depending on a more denotative meaning? Find examples from the text to support your answer.

"SHOULD ENGLISH BE THE LAW?" ROBERT D. KING

For Openers

Robert King gives the reader an abundance of examples of how "most nationalistic movements [have] a linguistic grievance." Do the many references persuade you? Consider how dominant American dialects separate regions: western, Texan, southern, New Yorker,

Bostonian, etc. Ask students for examples of how different speech patterns elicit positive and negative responses in various communities.

Teaching Strategies

1. Students may have difficulty reading this long essay. Help students track the argument and evaluate the evidence. Show how paragraphs are built with topic sentences, examples or an illustration, and a closing statement. Using this pattern, ask students to read the first line of each paragraph on a page, then read the whole page.

2. Bring post-it notes to class. Ask students to note their questions and observations as they re-read the essay by sticking post-it notes right next to the places to which they need to pay closest attention.

Collaborative Activities

1. Divide the essay so that each person reviews about 8 paragraphs. Have students read through the paragraphs, noting topic sentences. In order to appreciate and understand how the argument is sustained, have each person describe for the group the logical pattern that is developing section by section. Ask students to note how information is presented and interpreted and how it leads to a persuasive conclusion. Ask students to share post-it noted passages to see if they can answer each other's questions.

2. Ask students to list vocabulary words associated with dance, music, computer technology, and socializing. Ask them to interview people of different ages and from different locations to add to the list. Review the list to see how vocabulary changes. Discuss how possible it would be to try to legislate or prescribe the American language.

Writer's Options

1. Write a parody of the idea that laws can be passed to create unity. For example, write about a school where some outward sign or standard (like uniforms) is supposed to create school unity, where unity is actually created by something less tangible.

2. Describe a community to which you belong or one with which you are very familiar. . This group can be a club, a church, a family, a business, for example. Argue for or against King's point that diversity within a group can intensify and enrich the united whole. Consider what binds the group and how diversity gives it strength.

Multimedia Resources

Search the Internet for the "English Only" topics.   Look at http://www.mea.org/society/engonly.html

Suggested Answers for Responding to Reading Questions

1. If we accept the assumption that language binds and identifies a nation, then it follows that a nation becomes fractionalized by bilingualism. A particular language becomes the language of power, particularly in regard to economic control, causing a split in the society. The danger is social injustice and a lack of identity with the whole nationality.

   King cites Canada as an example of language diversity causing splits in a nation. He also refers to Russia and Slovakia and the Balkans.

2. In paragraph 26, King claims "a union transcending language" in reference to India's history with language diversity. He also refers to France, Italy, and Germany. He hold that traditions and a common history bind the nation. The comparison is valid in several ways. Individual states were colonized by particular interest groups and ethnic groups. States do seek to maintain their own identifiable characteristics based on particular traditions. However, the diversity in the language is not as extreme as it is in Switzerland, for example, where four distinct languages are spoken. All the dialects in the United States are considered part of English (or American).

   Whether King's argument is convincing is a matter of opinion and debate.

3. Although Amselle does not address this issue directly, we can extrapolate. Amselle is opposed to bilingual education programs because he sees learning English as the primary means of entering the mainstream and becoming an American. He would say that not knowing English bestows underclass status on immigrant children. For Amselle, becoming an American means learning English as quickly as possible.

Additional Questions for Responding to Reading

1. Look up "nationalism" in the *Oxford English Dictionary* . What other shared behaviors make up our national character? According to Amselle, how is language the most significant factor to national culture?

2. King writes, "History teaches a plain lesson about language and governments..." How familiar were you with the different historical references made in this essay? What impact have these references had on your understanding of the issue and have you been convinced?

"¡INGLÉS, SI!" JORGE AMSELLE

For Openers

1. List the different organizations and groups involved in the education argument (parents, teachers' organizations, NABE, Members of Congress). Consider who is missing from this list. For example, linguists are not included. Parents care for their children, but can they make the best educational decisions?

2. What additional studies should be made? For example, what is the long-term outcome of bilingual education? How is language best acquired? Are there psychological implications?

Teaching Strategies

1. Data is good evidence; however it is important to interpret data, not just place it in an essay. Amselle cites percentages and numbers throughout his essay. Discuss the importance or irrelevance of his numerical evidence. How are the statistics useful or irrelevant?

2. Amselle dismisses the research of Professor Collier as inconclusive because it has not been "subjected to peer review." Explain the concept of peer review to the class. Then look at the kinds of evidence Amselle brings to bear in his essay, and discuss its validity and effectiveness.

3. Amselle maintains a friendly and conversational tone throughout the article. Mark transitional devices and word choice that establish this tone.

Collaborative Activity

Ask students to discuss the importance of parental involvement in education. Ask them to share experiences from their own education in order to decide the wisdom of ignoring parental demands.

Writer's Options

1. Write a statement of philosophy about your responsibility as a voter and as an educated member of your community. To what degree are specific community problems the concern of the entire community?

2. Argue whether the federal government should or should not have the final decision on educational policies and practices.

Multimedia Resources

Look through the *National Review* for September 30, 1996. What other issues are discussed in the magazine? Does the magazine show a bias of any sort? Do you consider it a credible source? Look at articles in *The New York Times* and other national publications from this same time period. Based on the number of articles printed how important is the issue of language?

Suggested Answers for Responding to Reading Questions

1. Amselle uses statistics and news reports on how parents have reacted to schools' decisions to teach English. He could offer more theoretical support rather than numbers. Since there are no reference points with which to compare the numbers, they have little meaning.

2. If the federal government is listening to experts in the field, they may ignore the parents. The government may also be subject to pressure from state legislatures, lobbyists, or special interest groups whose votes are important.

3. One of Amselle's major points is that the federal government should reform bilingual education. In his essay, King makes the point that government policies can do little to change behavior--especially concerning language. He also makes the point that no matter what the government or educational professionals do, "not many of today's immigrants will see the first language survive into the second generation"(36).

Additional Questions for Responding to Reading

1. What is your own experience with bilingual education? How does your experience influence your reading of this essay?

2. If you permanently change your residence to another country where English is not spoken primarily, what language would you speak at home? Why?

CHAPTER 4

THE MEDIA'S MESSAGE

Setting Up the Unit:

Using Definitions to Clarify Your Writing

In many composition classrooms, students study and analyze the various forms of media with which they interact daily. These assignments are popular with students for several reasons: They can see how the media relate to their lives; the subject matter is inherently interesting; and students can often show their expertise in a particular genre, program, or medium. Instructors value this line of inquiry because it requires that students examine the everyday forces in their lives; that they consider how their viewing habits reinforce social and political views; and that they articulate possibilities for resisting the images they choose to reject. The resulting papers can be a refreshing combination of analysis and experience, of description and calls to action. For those interested in using a cultural studies approach in their composition courses, this chapter will provide some of the tools and perspectives for cultural critique.

As you and your students read through the essays in this chapter, you will probably notice that many of the essays depend on a particular term, which the writers define clearly. For example, in the very first essay of the chapter, "Television: The Plug-In Drug," Marie Winn clearly delineates what she means by *ritual*. In the next essay, "Giving Saturday Morning Some Slack," Charles McGrath defends cartoons by claiming that they are part of a childhood ritual. The term *ritual* falls under two perspectives. These first two readings establish the value of clear definitions, and the rest of the chapter illustrates what can happen when terms are—or are not—clearly defined.

Defining key concepts becomes an especially important strategy in public discussions of *media*, a term that in and of itself is often ill-defined. Journalists, talk show hosts, new anchors, and the like often use the word *media* to mean a nebulous group of forces that bombard the general public (another hazy term) in our homes, cars, and workplaces—while all too often excepting themselves from this group. Student writers often pick up this ambiguous usage and work in into their papers without considering the word *media*. In many ways, the very phrase "the media" has become meaningless. Your students will need to be wary of it.

If you choose to assign a paper based on the topics in this chapter, it is important to stress the importance of both definition and specificity. Ask students to keep a running list (perhaps in their journals) of the terms that are crucial to the writer's argument, as well as those words or phrases that seem important to go unexamined. They can work as a class or individually to arrive at working definitions for these concepts, which they can then use in their own papers if it is appropriate. It is certainly possible that different students might reach divergent understandings of a term, and this will offer an opportunity for the class to observe and analyze the ways in which words and meanings can shift.

Another potential rhetorical strategy for approaching this chapter would be to analyze the different types of argumentation styles used in the various essays. Many have similar themes, but most use alternative organizational plans and language choices.

Confronting the Issues

Option 1: Constructing Contexts

Many times we participate in a media event without knowing it. Since some kind of media permeates every aspect of our lives, it would be useful to have students demonstrate for themselves how much and how often they are in contact with the different forms the media can take. Before reading any of the selections in this chapter, brainstorm with your students to create a list of all the stimuli that can be considered a part of "the media." Your list might include:

- News Programs
- Newspapers
- Soap operas
- Television advertisements
- Talk radio
- Cartoons
- Comics
- Billboards
- Music radio
- Magazines
- Music videos
- Infomercials
- Sitcoms
- Talk Shows
- Reality TV

At this stage, students might argue about what constitutes "the media" and what does not. Agree upon a definition that your class can use. Once you have an inclusive list, ask students to record what sorts of media they notice, and the time they spend observing each kind. At the end of the week, total the types of media and average the times for each. Students can use this information in their papers, and they can refer to it as they read the essays in this chapter.

Option 2: Community Involvement

As part of a community outreach assignment, students can interview a range of people in the community (an arbitrary requirement would be for each student to conduct five brief interviews) to determine what types of programming people would like to have. Students can simultaneously research other types of programs that are available through public television or other local access stations. Draft a collaborative class letter that would go to a local affiliate of a national station, suggesting specific changes. Or use the findings from the class research in a collaborative letter to Congress regarding federal funding of public television. The "Student Voices" that begin this chapter provide ideas and further data.

Option 3: Cultural Critique

Tape at random a thirty-minute segment of MTV or VH1 and show it to the class. As a class, figure out how the time is divided. How much is advertising? How much is talk? How much time is spent on the videos? At the same time, ask students to catalog and describe the images they see of women, children, elderly people, African-Americans, Asian-Americans, Native Americans, people from other countries, animals, religious symbols, gay men and lesbians, factory workers, rich people, poverty-stricken people, urban settings, farms, suburbs, and so on. Continue to refer to this tape as you discuss the readings in the chapter, and discuss with your students the implications of these video clips in light of each reading. If you prefer, you can do this assignment with a local news program.

Option 4: Feature Film

Obviously, a chapter on the media offers many multimedia opportunities. Feel free to let your students help decide what direction you might take. One clear way to round out the offerings in this chapter would be to provide some historical context with the movie *Quiz Show*. This film depicts the budding game show industry in the 1950s, and comments about how audience reaction helped determine the outcome of the contest. Hailed in its time as the "brain-buster" game show, *Twenty-One* was, in fact, fully staged, choreographed, and rigged. When the scandal broke, millions of early television viewers felt betrayed; they mistakenly believed that what they had seen on television was real. The movie is long, and at times slow, but well worth the time and effort because of the issues it examines and the ways in which both the audience and the participants can be interpreted. If this movie is too long, consider showing instead the PBS show about the scandals.

Teaching "Two Perspectives"

In the two essays that follow, Marie Winn and Charles McGrath appear to take opposite sides on the issue of how television affects viewers. Ask students to write down their own responses before reading Winn and McGrath to the issue of how television affects viewers; ask them to revise those responses after they have read both arguments. Finally, students can discuss among themselves where they feel the blame should be placed; add to this picture as students work through the rest of the chapter.

"TELEVISION: THE PLUG-IN DRUG," MARIE WINN

For Openers

Winn discusses throughout this essay, but particularly in paragraph 35, the blurred lines between television and reality. Recently, numerous "real life" emergency shows have been created that further blur the distinction between "real" and "TV." Discuss in class why this might be. Do any of your students watch them? What is their appeal? How "real" are they? What are the positive effects they are supposed to have? What are the negatives? What might Winn say about these shows?

Teaching Strategy

Winn raises a number of questions about the assumptions people make about television, particularly those who feel television is a "neutral" technology. For example, in paragraph 8, she questions the role that television executives play in claims that television acts to bring families together; she shows how the mother in paragraph 12 assumes that watching *Zoom* is better than fixing dinner together; and she criticizes the mother in paragraph 14 for not seeing "anything amiss with watching programs just for the sake of watching." Are there other places you can find where the underlying assumptions made by Winn, or by the people in her examples, are questionable?

Collaborative Activity

Winn describes in paragraphs 21 through 23 the ways in which television has taken over the family gatherings. In groups, create lists of alternatives to television. Imagine the next holiday without television; what else would people actually enjoy doing? As a class, try to come up with a list that might appeal to the different generations in a family. Vote to identify the best three ideas.

What is the role of television in your family? During what hours is it on? What kinds of programs are on? Do family members watch together? What family activities are determined in some way by the television and its programming? Were any restrictions placed on your television viewing habits? Keep a log of which programs you watch, and for how many hours you watch them.

Multimedia Resources

Bring in some ads for TV sets from 1950s *Life Magazine* or other old ads depicting the happy family Winn describes in paragraph 5. Quite often these ads seem funny to students born in the late 1970s; ask them to figure out why. Then, compare these magazine advertisements with commercials for television today. What are the differences? What do the differences show about how times have changed in forty years?

"GIVING SATURDAY MORNING SOME SLACK," CHARLES MCGRATH

For Openers

Ask students to list the various Saturday rituals from their childhood and ones they now have at home and at school. Discuss other ways that Saturday distinguishes itself from other days of the week.

Teaching Strategy

The essay divides into three parts. First, McGrath describes his childhood Saturday morning rituals. Next, he moves to present day by describing television shows and schedules. Finally, he compares the two different experiences. Point out how unity is achieved by the cross-referencing of part two and three with part one.

Collaborative Activity

Bring in the television listings for the week, the entertainment section from the local weekly newspaper, and a Saturday newspaper. Ask groups to review one of these sources. What is happening locally on Saturdays? Do the TV programs or local events tend to divide families or are there programs or events that are opportunities for families to spend time together?

Writer's Options

1. Spend your own Saturday morning watching television so that you can support or refute McGrath's observations on the frenetic and rushed experience Saturday morning viewing suggests.

2. Write an essay that could be titled, "How to spend Saturday morning" or "A Lesson in Boredom: Saturday." Explain why your plan is an important part of the week's activities.

Multimedia Resources

Tape a set of Saturday morning cartoons including commercials and any introductions by announcers. What is the nature of the heroes in these shows? Is there evidence to show that the action does not form much of a plot, as McGrath suggests?

Two Perspectives: Suggested Answers for Responding to Reading Questions

1. Winn describes contemporary family life as individuals isolated from one another, with little active interaction or stimulation-- even though they might all be sitting around the same set. Television provides the structure for family time, the content for family discussions, and the shape of day-to-day existence. The television acts as a distraction and entertainment which competes with conflict resolution and dialogue.

2. Winn may agree with McGrath that TV watching is a ritual and that it provides down-time in an otherwise busy week. However, Winn is objecting to continual TV watching while McGrath is talking about Saturday mornings only.

3. Answers will vary. Students may point out that stories viewed on television may teach social responsibility, personal responsibility, vocabulary and other language skills, and conflict resolution skills.

4. McGrath finds commercials to still be what the 1950s and present-day cartoon shows have in common. Animal antics on cartoon shows continue to entertain. However, the animation of present-day cartoons is more sophisticated, and the story lines

are more complex. McGrath points out some serious absence in contemporary cartoons – for example, the absence of amiable adult figures and the absence of quiet moments of music and slow-paced plots. McGrath points out "there is seldom a quiet or reflective moment."

5. McGrath suggests that cartoons are a satire on personalities and human behavior. Winn sees television as isolating and insulating.

6. McGrath points out that television has very little positive representation of adults on its Saturday programs. He also points out that without the leisure of Saturday morning for both children and parents there is no break to the hurried life of the rest of the week. Television offers the possibility for relaxation, reflection, and quiet.

Using Specific Readings

"THE TRIUMPH OF THE YELL," DEBORAH TANNEN

For Openers

Tannen paints a very bleak picture of what she calls a "culture of critique." In class, define what she means by this phrase and discuss whether or not looking for opposing points inevitably leads to simplistic and erroneous arguments, as Tannen seems to indicate.

Teaching Strategy

Although she doesn't explicitly say so, Tannen makes the distinction between pathos and logos in paragraph 2; she characterizes it as the difference between "making" an argument and "building" one. Use this as an opportunity to explain the difference between logical appeals and emotional ones.

Collaborative Activity

Many of the problems Tannen identifies with the "two-sides of very issue" fight (described beginning in paragraph 6) are quite consistent with the American legal system. In groups, ask students to take a part of the essay and locate those places where Tannen describes behavior or argumentative strategies that would make perfect sense in a courtroom. Discuss the implications of the situation. Could Tannen's criticism also apply to our judicial system?

Writer's Options

Write about a time when the dynamics of an argument caused you to take a more extreme position than you really felt. What happened? What were the consequences of your position? If you could have the discussion over again, what would you do differently?

Multimedia Resources

Tannen appears on talk shows quite frequently, yet she seems surprised when she describes (in paragraph 14) the dynamics of the argument. Bring in a tape of a recent episode of the many talk shows and discuss the conventions within which they work. As an expert in communication dynamics, is it possible that Tannen expected something different? What might she have been able to do to change the course of events?

Suggested Answers for Responding to Reading Questions

1. As she claims at the beginning, when people fight, they often do not listen to each other. Matters of public policy are too important to be decided without seriously listening and considering possible implications, which are more often than not lost in the fighting.

2. Tannen believes that reducing complex arguments to two sides obscures important points and leads to faulty and erroneous conclusions; likewise, looking for sides where there are none only creates unnecessary arguments among questionable "authorities." As she says in paragraph 7, "[t]ruth is more likely to be found in the complex middle than in the simplified extremes."

3. Answers will vary; more than likely students will cite the potentially competitive nature of graduate school.

Additional Questions for Responding to Reading

1. Do you agree with Tannen's characterization of our "culture of critique"? If so, why? If not, how would you describe American public debate?

2. In paragraph 16, Tannen warns against "modeling intellectual interchange as a fight." What might be other possible models? do you feel that this tendency trickles down into other aspects of our lives? If so, explain how.

"SEX, LIES, AND ADVERTISING," GLORIA STEINEM

For Openers

Bring in several *Ms.* magazines. Also bring in copies of *GQ* and *Vogue*. Find examples to illustrate Steinem's references. For example, look at ads, feature articles, and letters to the editor and from the editor.

Teaching Strategies

1. Clarify the main argument of the essay. *MS.* magazine needs ads in order to stay in business. However, they want to maintain a particular ethical standard for their magazine. Determine what Steinem's essay sets up as the criteria for an ethical standard for her magazine. By appealing to lesbian subscribers, is *Ms.* making an ethical or a business decision?

2. Create a four-column grid on the board or on an overhead. As a class, list the different possible advertisements in the first column. In the second column, list the objection the magazine has with the product. The third column describes the product's company's objection to the magazine. Using this grid, look for a pattern and see if that pattern supports Steinem's claim.

3. Discuss how the combination of narrative and argument work together. What effect does Steinem's use of bulleted paragraphs have on your ease of reading and on the tone of the article? When might using bulleted paragraphs work in your own writing?

Collaborative Activity

First, divide the class into groups of between 3 to 5 students, depending on the size of your class. Assign a part of the essay to each group. In part one, Steinem writes about the problem of attracting advertisements. In part two, she discusses the relationship of magazine essays to its audience. Part three proposes a "what if?" and states her great frustration. Then, ask each group to look at different magazines in terms of Steinem's specific objectives and objections to magazine publication. Look for evidence that supports or refutes her views.

Writer's Options

1. Steinem writes from an insider's vantage point as a publisher. She gives enough information about the magazine and advertising industry to make a credible argument. As a consumer, or an outsider, choose one of your favorite magazines. Are the advertisements in the magazine set up in ways that Steinem describes? Are there "advertorials"? If all the advertisements were eliminated, would you pay more for the magazine? To what degree do magazines and their advertising influence your personal life?

2. Steinem is opposed to the objectification and stereotyping of women in advertisements. Analyze two or three magazines or two or three different kinds of products to refute or defend her view of most advertisements. Then write an editorial for your school newspaper that discusses the extent to which you and maybe your audience depend on magazines for information and for shaping your views on social issues.

Multimedia Resources

Clothing and cars are heavily advertised. Contrast the ads in magazines to ads on television. Ask students to record a favorite TV advertisement and bring it to class. Be sure they note the program associated with the ad and the time it came on the air. Advertising kind of advertisement to see if it supports Steinem's views.

Suggested Answers for Responding to Reading Questions

1. For discussion. Answers will vary.

2. Consider reasons for negative or positive answers.

3. Answers will vary.

Additional Questions for Responding to Reading

1. Steinem's essay uses bullets to mark sets of paragraphs. Why are those sections set apart from the rest of the essay?

2. List the negative effects advertisements can have on magazines, according to Steinem.

For Openers

Bring in several issues of the *New Republic* to class. Look at the types of articles and advertisements and read letters to the editor. Who would you say is the subscribing audience of this publication? Discuss why your students would or would not subscribe to this particular publication.

Consider the connotations of the title. Explore possible connections to fishing and to Captain Hook in *Peter Pan*.

Teaching Strategies

1. Crouch makes metaphorical references to fishing and whaling throughout his essay. Ask one or two students to list these references on the board as other students identify them. Then discuss how Crouch's language is effective or unnecessarily inflated.

2. Ask the class to list Crouch's objections to Jackson's work. Are these objections supported by evidence? What is the quality of that evidence?

3. List other public figures to which Crouch might also object, such as sports figures, politicians, or celebrities. Dennis Rodman, Michael Jordan, Bill Clinton, Ted Kennedy, Elton John, and the like are examples.

Collaborative Activities

1. Ask students to assemble various recordings and videos by Jackson; choose examples that show Crouch's objections to Jackson's popularity, and in a multimedia presentation demonstrate those objections.

2. Have students poll the members of their group to find out to what performers they prefer. Based on the criteria set by Crouch for popularity in the music industry, would these other performers be considered artists by Crouch? What is their response?

Writer's Options

1. Argue for the worth of Michael Jackson's song lyrics by analyzing four of his songs for quality of verse and thematic value.

2. Choose song lyrics from different parts of Jackson's career. Identify changes, patterns, and developments in his career as a performer. Research interviews with Jackson to support or refute Crouch's claim.

Multimedia Resources

1. Visit the Web site: www.mjam.com for links to different images and information on Michael Jackson. Compare the biased or loaded words in Crouch's essay to those used in this Web page.

2. Visit the Michael Jackson fan club at http://fred.net/mjj. Does this fan page support Crouch's view of the followers of Michael Jackson?

Suggested Answers for Responding to Reading Questions

1. Because everyone knows Michael Jackson, and because he is such an extreme case, one can see in him how identity, family, and art are sacrificed for the lure of money and fame. Crouch senses that Jackson sends the negative message of self-dissatisfaction, loss of family, and loss of values. What is gained in this loss is a superficial fascination with glitter and wealth. Agreement is a matter for discussion.

2. Because the metaphor of fishing is used as a constant reference in the essay, the title acts as a unifying agent to the essay. Titles such as "White Glove Treatment" or "Pop and Jello," may have been effective also, but the metaphor would have to change from the fishing images to something else. The title "Hooked" refers to the opening line in journalism as the "hook" to get the reader quickly involved in the article so that they will read on. It also refers to catching a prey so that it cannot escape. In addition, "hook" connotes the character of Captain Hook from *Peter Pan*.

3. Answers will vary.

1. How does the opening paragraph set the tone for the essay? What relationship is established immediately between Crouch and his reader?

2. Re-read paragraphs 3 and 4. How does sentence structure create emphasis and rhythm? Notice in particular the length of phrases in a sentence and sentence length. Are there any outstanding words or phrases that you consider particularly effective or ineffective?

"THE MOVIE THAT CHANGED MY LIFE," TERRY MCMILLAN

For Openers

Ask students with which character they identify most in *The Wizard of Oz*, and on the blackboard keep track of the different responses. How different are reasons for students who choose the same character? Have students favorite characters changed as they grew older? Why might certain characters be more appropriate at different stages of life?

Teaching Strategy

How many of McMillan's reactions are different from those of your students? Which of her interpretations in the first half of the essay are surprising? For instance, how many students had perceived Auntie Em as nasty? How many knew on their first viewing that Dorothy was only dreaming?

Collaborative Activity

In groups, find the places in which McMillan uses the word *happy*. Each group can take a different section of the essay. What does happy seem to mean each time it appears? Does its meaning change throughout the piece?

Writer's Options

McMillan wrote this essay for a collection on movies that changed writers' lives. What movie changed your life? Write about it, following the significance of the plot as it unfolds, just as McMillan does.

Multimedia Resources

In a non-musical Disney-made sequel to *The Wizard of Oz* called *Return to Oz*, Dorothy is institutionalized because of her persistent delusions about her time in Oz. In many ways, the Oz in the sequel is closer to the Oz that McMillan describes at the beginning of her essay. View the film with your students, and ask them to describe how McMillan's life might have been changed had she viewed the sequel. What lessons, values, and characteristics would be different?

Suggested Answers for Responding to Reading Questions

1. McMillan's life in Port Huron was like Dorothy's life in Kansas in several important ways: She too, wanted to escape the dreariness of her life, to find a place where there was no trouble, to find happy people, and to find a bit of power for herself in a life dominated by an Auntie Em. These similarities mattered much more than the differences-- a blue-collar world rather than a farming community, a large, noisy family rather than a one-child family, and African-American family rather than a white one.

2. Dorothy eventually loses her appeal for McMillan because of her desire to return to the life that she left, the life McMillan wanted to be able to leave. When presented with infinite possibilities, Dorothy returns to Kansas with a resolve to be happy with what she has, rather than trying to change it.

3. Answers will vary.

Additional Questions for Responding to Reading

1. Analyze the structure of this essay. How many sections are there? Why might McMillan have chosen to arrange her essay as she did?

2. McMillan offers some alternative endings to the movie. What alternative endings can you suggest? Which would you prefer?

For Openers

If you feel daring, ask a group of students to use a karioke machine for class — or have them do a lip synch of a fifties' song, complete with synchronized dancing.

Teaching Strategies

1. Trace the words and phrases in the essay that suggest women's feelings of powerlessness and silence--for example, "own victimization" and "tap of history." Divide the paragraphs among the students. Ask them to underline negative words and phrases. Discuss how the preponderance of negative terms intensifies the claim that this music was important to the sexual awareness (and revolution) of women of this generation.

2. Is this essay dated, or is it still true that "for answers -- real answers -- many of us turn to the record players, radios, and jukeboxes of America" (6)? How does technology affect today's relationship with music? Take into consideration the high tech productions of MTV, CDs, portable cassettes, and videos.

3. The author uses active voice effectively throughout the essay. Transfer a paragraph to an overhead transparency. Go through the paragraph with the class, changing active voice to passive voice. Demonstrate how the energy and immediacy of the presentation is lost.

Collaborative Activities

Divide the class into groups with the purpose of studying the cultural environment in which these songs were written.

1. Find popular magazines from the late 50's and early 60's. How are women and girls depicted in advertisements?

2. Look at the various issues covered in the Detroit and New York newspapers from the 50s and 60s. What is important? How many women's issues were covered or raised?

3. Check out movies produced during the 50s and 60s. What are the titles? How do these movies describe the culture?

4. Look at news accounts and trace the Civil Rights Movement. What is the state of affairs during the late 50s and early 60s?

5. Ask each group to report to the rest of the class. In view of the cultural environment, does the urgency of the songs seem more acceptable?

Writer's Options

1. Could an argument be made that a particular type of music in the 90s gives voice to an otherwise silent group? Consider rap, ballads, and jazz. Focus on a particular group that is well-represented, but feel free to refer to other groups as well.

2. Choose four or five songs from the 70s, 80s, or 90s. Analyze how relationships are represented. Make a claim that these songs help a group find its identity and morals. Research what social issues were concurrently pressing social groups.

Multimedia Resources

Have students watch any movie with Pat Boone or Elvis Presley. Ask them to note in particular the roles women played and how they were treated. Compare a movie with Pat Boone, such as *Bernadine*, to a present-day music video by Janet Jackson. The movie, *Where the Boys Are*, could also serve this same purpose.

Suggested Answers for Responding to Reading Questions

1. According to Douglas, the music of the Shirelles, and of others like them, began to give a voice to women's issues and identity.

2. Douglas cites specific titles that describe, in many cases, the song's theme. She then picks one song and summarizes the story-line of the song to show how messages were implied. She does this throughout the essay, making her references specific to an audience who does not need to be totally familiar with the original music.

3. A matter of opinion. In answering this question, identify the target audience of this essay.

Additional Questions for Responding to Reading

1. The author provides many examples of different songs. What effect do all these examples have on readers?

2. Douglas's enthusiasm for her topic is evident throughout the essay but her enthusiasm does not overshadow her research or the logic of her argument. Find places where her presence is evident.. Discuss how enthusiasm for a topic affects the writing process and the product.

"WE TALK, YOU LISTEN," VINE DELORIA, JR.

For Openers

The second language issue Deloria describes beginning in paragraph 3 is still contentious today (see Chapter 3 on language). Ask your students to describe ways in which they see minority groups being marked by their language use. Deloria calls this marking the first aspect of stereotyping; do your students agree? Is it necessarily negative?

Teaching Strategy

Deloria creates a strange tension between building solidarity with and keeping distance from other minority groups, particularly Hispanics and African-Americans, but he does this in order to show that Native Americans are still in the worst position. Discuss with your students how Deloria does this—for example, how he sets up each part of his argument with descriptions of other groups first. Does the strategy work?

Collaborative Activity

Each group can take a different segment of American society and read Deloria's piece from that perspective. To set up this activity ask the entire class to list on the blackboard various groups that might be interested in responding to Deloria. Students will then need to try putting themselves in somebody else's shoes to answer Deloria from a non–Native American position.

Writer's Options

To create a powerful stylistic effect, Deloria uses repetition and contrast in his descriptions of his people. In paragraph 1 he mentions "real Indians with real problems." Later, in paragraph 12, he says that it is the "strangeness of Indians that made them visible, not their humanity." In the first case, the placement of the repeated word strengthens the sentence; in the second one, the contrasting word creates an emphatic sentence. Try rewriting some sentences you wrote in an earlier entry following these two forms of Deloria's. Use them in your next paper if you like the way they work.

Multimedia Resources

Bring in scenes from the recent Disney film Pocahontas to see if Deloria's 1970s concerns still ring true. Let students analyze the film and decide for themselves.

Suggested Answers for Responding to Reading Questions

1. One of the main purposes of the ethnic studies programs is to educate people to see beyond media stereotypes. On the other hand, Deloria shows that the programs can, themselves, create new simplistic views of ethnic groups.

2. Answers will vary, but students will be able to point to certain movies and television programs, including *500 Nations*, *Dances with Wolves*, *Geronimo* and *Northern Exposure*.

3. In his focus on stereotypes, Deloria didn't notice that many groups simply aren't depicted at all, either positively or negatively. He simply doesn't mention them. He also excludes women, gays and lesbians, and a host of other ethnicities such as those of Asian descent.

Additional Questions for Responding to Reading

1. What examples can you give in today's media of the "patriot chief" interpretations of history (paragraph 19); the "cameo school of ethnic pride" (paragraph 22); and the "contributions school" of middle class materialism (paragraph 24)? Are any of these representations more prevalent than the others? If so, why?

2. In the last paragraph, Deloria says that the "problem of stereotyping is not so much a racial problem as it is a problem of limited knowledge and perspective." What can be done to solve this problem?

For Openers

Ask the class to create a list of the characteristics they most associate with adolescent boys, drawing on Stark's essay as well as their own experience. What attributes can they add to Stark's own list? How do their qualities match with the examples that Stark's selects? Do they continue to bolster Stark's argument or do they contradict it?

Teaching Strategy

In paragraph 16 Stark indicates that networks and manufacturers have a financial stake in promoting a culture that embraces the values of adolescent males. Discuss this assertion and its implications.

Collaborative Activity

Stark uses a wide variety of television genres, products, and activities to show the prevalent preoccupation with preadolescent values. In groups, locate the types of examples Stark uses, and provide counterexamples within each type. To whom does each example appeal? Why are both the examples and the counterexamples so popular?

Writer's Options

In paragraph 8 Stark claims that "much of journalism today is really a form of institutionalized early adolescence." Write about an example that would support Stark's argument. If you followed the O.J. Simpson trial, were there times when you felt it was like this? If you listen to talk radio, can you cite some particularly adolescent examples? Write a description of such an event, and discuss why you felt it was adolescent.

Multimedia Resources

Any number of clips from television shows would work well here. Ask your students to watch for good examples in their own viewing. If any have access to VCRs (some may have them in their dorms these days), ask them to tape something and bring it in to share with the rest of the class.

1. Answers will vary.

2. Answers will vary, but examples might include TV shows like *Ally McBeal*, and *The Oprah Winfrey Show*; or movies such as *Thelma and Louise* and *Fried Green Tomatoes*, which bills itself as representing "every woman."

3. Answers will vary.

Additional Questions for Responding to Reading

1. Would you react to this piece differently if a woman wrote it instead of by a man? Explain your answer.

2. In paragraph 13 Stark writes "That propensity to be in the world of sex but not of it is certainly a sign of the times." What does he mean by this? What is the difference?

"TESTIFYING: TELEVISION," WENDY KAMINER

For Openers

1. Ask students to freewrite on the cliché: "Confession is good for the soul." Discuss the results.

2. If viewers do household chores, eat, and socialize while watching television, how seriously should they be expected to listen to the talk shows?

Teaching Strategies

1. One way of understanding a text is to examine its rhetoric. Ask students to list opposing terms as they read this essay. For example, in paragraph seven, the word *listen* contrasts the word *talking, self-absorption* contrasts *self-expression*. After creating the list of opposing words and phrases discuss the details of Kaminer's claim.

2. Kaminer is quite self-assured in her essay. Is she credible? If so, how does she establish that credibility with her audience? If she is not credible, what does the essay need?

Collaborative Activity

Kaminer uses humor in several parts of her essay ("The family that reveals together congeals together."). What effect does such humor have on the reader? Is she trying to make us feel good about her essay? Divide the essay into sections and assign a group to each section. Ask student groups to find humor in their section of the essay and to highlight it. What is the role of humor to the argument? Are there sections where there is more or less humor? Consider if humor is strategically placed.

Writer's Options

If television serves a role beyond entertainment, then what obligations does it have in providing instructional examples for conflict resolution, cultural development, or any other personal skill?

Multimedia Resources

Kaminer states that the guests on talk shows are not particularly articulate and that "they rarely finish their sentences." Bring in two television clips: the first one should be a ten-minute clip of a talk show; the second should be a ten-minute clip of a television drama. Ask students to count the number of complete sentences and to observe the diction, level of vocabulary, and use of clichés. Does their evidence support or contradict Kaminer's opinion?

Suggested Answers for Responding to Reading Questions

1. Kaminer sees confessing as personally reflective. In telling one's story or faults, one is talking to one's self for the benefit of one's self improvement or awareness. Testifying is meant to be public, for the public edification, information, and judgment.

2. Answers my vary. Students may want to research how these guests are paid.

3. Answers may vary. Be sure student support their opinions with specific evidence.

Additional Questions for Responding to Reading

1. Using a dictionary, look up the meaning of the words *solipsism, voyeurs, codependency, impugn, and perspicacity.* Then find

synonyms for these words in a thesaurus. Substitute the synonyms for the words Kaminer chose. What is lost or gained by the substitutions?

2. After reading Kaminer's biographical introduction, how are you influenced as a reader by her affiliations to different groups?

Focus: Does Media Violence Hurt?

The following essays question the right of the consumer to suppress or censor a film, TV program, or other media presentation on the grounds that it adversely affects the culture and degrades moral perspectives. Ask students to write about a movie, TV program, song, or video game that they found morally disturbing. Is being disturbed in this way necessarily harmful? Then, ask students to write a definition of "violence" and "entertainment." After reading the following essays, ask them to re-visit their definitions.

"UNNATURAL KILLERS," JOHN GRISHAM

For Openers

Make a chart on the board. Head the columns of this chart: "Type of Movie," "Name of Movie," and "Possible Influence on Viewer." Then ask students to fill in the columns. Types of movies might be "Action," "Romance," "Mystery," and "Horror." Your purpose is to determine the range of responses that movies might cause.

Teaching Strategy

Establish the criteria for satire. You can look up a definition in any literary handbook or you can use a dictionary definition. Satire renders a subject ridiculous and contemptible. What would a satire about mass murder have to do to evoke a satirical response? Ask students to suggest possible plots that would fit the criteria for satire.

Collaborative Activities

1. Ask groups to research the two murder cases in this essay. Compare other reports of the murders to Grisham's view of the case.

2. Based on the essay and additional research, enact a courtroom trial of the Ben and Sarah murder trial. Have groups prepare as attorneys for the defense, prosecutors, and TV reporters covering the case.

Writer's Options

1. Appealing to the emotions of your reader, write an essay reminding them of their power as consumers to boycott and sue industries that produce damaging products.

2. Some say that movies allow viewers to enter the dreamland of their own subconscious. Referring to the class discussion of various kinds of movies and their influence on the viewers, argue for or against a particular type of movie. Is our culture being shaped by a strong appeal to desire and dream? What conclusions are you willing to make?

Multimedia Resources

Show some clips from *Natural Born Killers* that could be labeled as gratuitous violence.
Contrast the movie *A Time to Kill* with *Natural Born Killers*. Both movies involve violence and murder. How are the effects different?

Suggested Answers for Responding to Reading Questions

1. The background information shows why Grisham felt sympathetic toward Mr. Savage. It helps make his relationship to the victim personal. A reader might suspect that Grisham is just interested in publicity, but the background information helps the reader better understand why Grisham was upset enough to write this article.

2. Grisham is obviously disgusted with the celebrated violence in the film *Natural Born Killers*. However, he is worried about the effect of this type of violence on our culture. Grisham would like to influence movie-going consumers to boycott violent movies, because until the movie industry begins losing money on such films, its motivation to discontinue them is minimal.

3. The evidence that there is a cause and effect link between viewing the movie and committing murder is thin. It is based on the testimony of the murderers who, it has been established

by Grisham, are seeking to blame their actions on something or somebody else.

Additional Questions for Responding to Reading

1. What would cause you to boycott an establishment or refuse to purchase a particular product? As a consumer, how do you perceive your power?

2. What other movie or form of entertainment do you think has potential to adversely affect our society?

"WHY BLAME TV?" JOHN LEONARD

For Openers

In contrast to the negative stories of the "trash" that regularly appears on television, Leonard provides some very positive examples of good programming. In class, ask students to create two lists of their own, one on each side of the board, of positive programming and trash. Next, figure out which channels or networks offer the most examples on each side of the board, and assess the implications of the availability of each kind of programming.

Teaching Strategies

Leonard uses a number of stylistic devices you may want to call to the attention of your students. Discuss their views on these techniques, and add your position on them, explaining to students how you would react if they used some of these strategies in their papers.

Repetition: "Never mind" runs through the first few paragraphs, and is echoed later; "or" is scattered throughout the essay, particularly in paragraph 5.

Parenthetical comments: Phrases, explanations, even full sentences appear in parentheses throughout the essay.

Questions: The piece ends with a series of items that are questions.

Collaborative Activity

Leonard relies on lists to illustrate his point throughout this essay. Ask groups of students to examine the lists that appear in a particular group of paragraphs, and characterize the nature of the item in the lists. Some include purposely trivial items; some include pairings that normally would not be linked; some simply build on themselves to show how many positive examples he can accumulate. Each group can report its findings to the class and explain what Leonard achieves with each kind of list he uses.

Writer's Options

Leonard uses a number of stylistic strategies to make his writing sound very conversational. In paragraph 2 he uses the word say in the middle of his explanation to show how arbitrary he thinks the decision was; in paragraph 4 he uses the nonstandard grammatical "the littlest firebird must have got his MTV elsewhere" to mimic the ad campaigns of that network. In your own journal, try rewriting something you have written so it will have a more conversational tone; put in some colloquialisms that would never be considered "proper" English. Used appropriately, they can enliven your writing.

Multimedia Resources

Bring in a photograph of Ernest Hollings, or a C-SPAN clip of the proceedings that Leonard describes in the first two paragraphs. Read aloud Leonard's opening description and discuss what he accomplishes with his imagery. How does it work (or not work) with the "real" thing? What other images might students replace the warrior-king with? What interpretations might result from substituting a different image?

Suggested Answers for Responding to Reading Questions

1. Answers will vary.

2. Answers will vary.

2. Leonard is making the point that children are not raised by television, communities raise them. Violence is not a product of television, according to Leonard, and we must look to other causes. His argument is logical because he challenges assumptions made about the effects of television. Grisham would not agree with Leonard's argument. His evidence leads to the conclusion that movies and television suggest criminal behavior to the point of causing it.

1. Underline words and phrases in the first four paragraphs that create the mocking tone and bias of the essay.

2. What concessions does Leonard make to an opposing point of view? Notice where he places these concessions in his essay.

"MEDIA AND THE ADOLESCENT," MADELINE LEVINE

For Openers

1. Discuss negative messages given to adolescents by parents, teachers, or society at large that make teenagers feel like "nobody."

2. Besides movies, where might teenagers receive affirmation of their own worth? Ask for actual examples.

Teaching Strategies

1. Debate whether movies should be rated more strictly so that adolescents cannot see violent movies.

2. Review the essay, and decide as a class if its assumptions are valid. For example, in paragraph 12, the essay reads, "No longer are parents repositories of wisdom, filling up eager young vessels with their knowledge, life experience, and sense of morality." Here the author is assuming that this is a new situation and that the previous generation acted as vessels to their parents' experiences. Does everyone agree with this assumption?

3. In the same paragraph, Levine states, "Teenagers are eager to construct an identity of their own, one far less dependent on their parents, and more in line with their own strengths and weaknesses." Does the class agree that this is a unique phenomenon of this generation, or do they believe it is a general condition of being a teenager?

4. Look for other assumptions in the essay that we as readers should carefully consider.

Collaborative Activity

Have students list all the movies seen in the past month by members of their group. Then, ask them to divide the movies into categories based on Levine's article: sex, violence, sex and violence, human interaction without violence, etc. What kinds of movies do they most watch? List favorite movies seen in the past year. Levine suggests that certain movies appeal to different genders. Does students' list of favorites support her claim?

Writer's Options

1. Levine suggests that adolescents "need significant amounts of time to 'do nothing.'" In a narrative essay, show how this use of time is important. Allow the reader of your narrative to hear the thoughts of an adolescent exploring "internal preoccupation" while engaging in activities that adults might see as a waste of time.

2. Create a new grading standard to rate movies (similar to our G, PG, PG-13, NC-17, R, and X ratings). Describe the criterion for each category. Finally, write an explanation and a defense of your new rating scale. Be sure to explain any assumptions or values reflected in your standard.

Multimedia Resources

Consider the characters in such movies as *Father of the Bride* (1991) with Steve Martin and Diane Keaton; *Mrs. Doubtfire* (1993) with Robin Williams; *The Natural* (1984) with Robert Redford; *Phenomenon* (1996) with John Travolta; and *Junior* (1994) with Arnold Swarzenegger and Danny DeVito as examples of heroic but non-violent males. Can students suggest others?

Suggested Answers for Responding to Reading Questions

1. "Evil has its attraction. . ." suggests human nature is at fault. "It is a peculiarity of our culture. . ." suggests that social values are at fault. "In spite of industry's claim. . ." points the finger at the movie industry's own moral corruption and greed.

2. Yes, Levine does contradict Grisham. Grisham claims mindless or misunderstood representations of violence in the media breed violence in reality.

3. Levine holds that media violence helps create a forum for abuse and murder. Not only is media violence a pervading influence, but it also takes away from time spent on more wholesome and creative mental engagements. Levine's solution is to create a

97

constant call for real role models and a respect of individual rights, especially on gender relationships. These must become inculcated in the family and in the larger society.

Additional Questions for Responding to Reading

1. Find places where Levine draws from theories of psychology to explain the effects of the media. Is the presentation of the theory clear and fairly applied?

2. According to this essay, how do psychologists gather evidence?

3. How can you determine who is the intended audience of this essay? Is the evidence gathered for the essay appropriate for that audience? Why or why not?

CHAPTER 5

WOMEN AND MEN

Setting Up the Unit: Using Gender-Inclusive Language
for Audience Awareness

The topic of this chapter, women and men, will undoubtedly spark some lively discussions. The readings in this chapter address a number of issues, from relationships between the genders, to relationships within a gender. Because students are aware of the impact each gender has upon the other, this chapter provides a good opportunity to discuss the conventions regarding gender-inclusive language. Many instructors like to state their policies about sexist language (if they have them) at the beginning of the term, but others let this subject arise naturally from their students' papers or discussions. For those choosing the latter course, this chapter's readings will give you the opportunity to raise concerns you might have as teacher.

Gender-Inclusive language, simply put, means language that does not exclude one gender or the other with word choice. More often than not, this designation refers to using male terms when describing both men and women; "mankind" is on example. Most instructors (and publishers, for that matter) now have policies that require inclusive language, but they often do not provide adequate instruction on how to avoid sexist constructions. In your discussion of gender and language, you will want to offer students some concrete strategies for replacing language they may have heard all of their lives with options that are more responsible. The most common strategies include:

- When appropriate, changing a subject to a plural form with the pronoun referring to many people rather than a person of a specific gender.
- Alternating between males and females in different examples.
- Using *he* or *she* or *him* or *her* rather than simply *he* or *him*.

99

A grammar handbook with a section on avoiding sexist usage can offer other suggestions.

When discussing these strategies in class, many instructors face considerable opposition from students. Many students truly believe that *he* is a generic term; other recognize the difficulty of revising well-established language patterns to accommodate women. Because student resistance to this issue can be so high, it is helpful to present gender-inclusive language as an audience issue. Since non-inclusive language is likely to offend, it makes more sense to avoid it. And, since gender-inclusive language does not offend anybody (people rarely notice it when it is used skillfully), it is the safest way to go. This strategy works well with today's market-minded students. If your class discussion on this subject reveals that students want more information, or more evidence, refer them to Alleen Pace Nilsen's "Sexism in English: A 1990s Update" found in Chapter 3.

Confronting the Issues

Option 1: Constructing Contexts

Present the class with each of the following scenarios, one at a time. In each case, ask students how they would react to the scenario, how they think most males would react, and how they think most females would react. In class discussion, find out if the students react as expected. Try to account for all the students in the class, those conforming to the stereotypes and those not conforming. Discuss, too, the differences in the expectation of gender behaviors and the potential problems involved in building those expectations.

- At a sporting event, someone insults your date.
- You are driving alone, and one of your tires blows out on a busy highway.
- It is 11 P.M. and you miss the last bus home; you live a mile away.
- Your dentist places a reassuring hand on your arm.
- Your best friend drops buy with his or her new baby.
- You are told by a recruiter that the job for which you are applying has traditionally gone to a man because it calls for supervision of an all-male staff.

- You go with a male friend to see a sentimental movie that really touches you emotionally.
- Alone at a fraternity party, you suddenly feel ill. You don't really know anybody at the party, but one of the fraternity brothers tells you that you can lie down in his room until you feel better.
- You see a young teenager harassing an elderly man.
- You have always been interested in the health professions, but you do not feel you can invest the time or money required for medical school. You learn you have been awarded a full-tuition scholarship to a four-year baccalaureate program in nursing.
- You are given tickets to the Superbowl and to the hottest new musical in Broadway. The only catch is that they are both for the same day.
- You read in your college newspaper that a student you know has filed rape charges against another student whom you also know.
- A teacher praises a literary work that portrays women in a negative—even insulting—manner.

After discussing the responses to these scenarios, and after reading through the chapter, return to these questions. Ask students if they would respond differently to any of these scenarios now. Which ones? How would they act differently? Why?

Option 2: Community Involvement

Find out what types of services and programs at your school address gender-related issues—for example, rape awareness programs, security escort services, women's studies programs, women's athletic programs, gay/lesbian/bisexual organizations, sexual harassment policies and guidelines, and so on. Search out any resources that might be available, and create a handbook for incoming or returning students.

Option 3: Cultural Critique

Ask students to bring in some of their favorite magazines, and ask them to describe the images of men and women they find in them. Are there common characteristics? How much variation is there? What sorts of things are men and women doing? What are they saying? What are they wearing? What are they selling? What parts of their bodies show? Students usually find gender norms quickly and easily in magazine ads.

Option 4: Feature Film

For some historical context, and for a popular film that covers many of the issues throughout this chapter, show *A League of Their Own*, a movie about the all-American girls' baseball league during World War II. As students watch, ask them to keep a list of problems that arise that have to do with gender. In discussion, ask them if they think these issues might be resolved differently now.

Teaching "Two Perspectives"

The two poems that follow capture an adult's view of children caught or avoiding the traps of conformity and gender stereotypes. Ask students to reflect on their own experiences with toys that helped create gender stereotypes. Also, ask them to describe some social situation where children engaged in power struggles based on strength or daring. Finally, ask students to discuss the risks involved in non-conformity. What literary or real figures are examples of non-conformity?

"BARBIE DOLL," MARGE PIERCY

For Openers

According to the headnote to the poem, Piercy wants to write "poems that speak to and for" her readers. Does she? Do your students feel that Piercy accomplishes this with "Barbie Doll"?

Teaching Strategies

1. It has been said that if Barbie were a real woman, her measurements would literally immobilize her. Point this out to your students, and find out what their images are of Barbie. What does (or did) Barbie represent to your students? Do they really believe that somebody can (or should) look like Barbie?

2. Consider the reasons why the Barbie doll is the most successful toy of all time. What do children (and their parents who buy it) get from this toy? Why is Barbie such a powerful cultural icon?

Collaborative Activity

In groups, ask students to write a proposal to the Mattel company suggesting the company either modify the Barbie doll or create a new toy. Proposals should include a section on the type of child (or parent) who would be likely to purchase this toy, as well as the reasons why it would be attractive. Include drawings if possible.

Writer's Options

Ask students to pick a toy that became an icon of their childhood, and write about why that toy was so important to them.

Multimedia Resources

Bring in an actual Barbie doll, along with pages from a department store catalog that shows Barbie and all of her accessories. In class, categorize the types of outfits she wears and the types of other merchandise that is sold along with the doll. What are the implications of the activities associated with Barbie? What are the implications of the huge industry that appears to exist solely for Barbie's leisure?

"RITES OF PASSAGE," SHARON OLDS

For Openers

Most cultures have some rite of passage to mark various stages of growth or designate roles in the community. For example, in the Moravian society, females wore different colored ribbons on their bonnets as signs of age, place in the community, and marital status. In Europe, a boy received long pants to acknowledge his becoming an adolescent. What are some rites of passage that we have in the United States? Some examples are getting a driver's license, applying to college, receiving a first paycheck, etc. Consider other smaller rites of passage that are also important.

Teaching Strategies

1. For as many students as there are in your class, select words from the poem and assign each student a word-- for example, *guest, party, gather, living room*, etc. Ask your students to spend some time listing all the meanings and associations they have for that word. Bring a dictionary or the OED to class to help students find multiple meanings to words such as *keel, jockeying, frown, turret*. Read the poem aloud. Then ask students to read their associations for the individual words. Read the poem aloud again. Discuss the different meanings and multiple levels of meaning that the poem now carries.

2. The mother-speaker of the poem obviously admires her son. Ask students to point out descriptions that depict her son in a favorable light. Consider how he is described as the "host," that he acts "for the sake of the group, "freckles," and the constant reference to the boy as "my son."

Collaborative Activity

1. Ask groups to consider why the boys want to fight. Is this a natural competitive response, or does this poem reinforce some of the essays' claims in Chapter 4? Do the media teach children this kind of behavior?

2. Olds sets up an interesting dichotomy between the "little men's" behavior and the son's. List examples of the different descriptions. Discuss the paradox that emerges. This paradox is created when the little men with all their grown up posing create conflict, but the little boy, described as a child, displays wisdom and maturity to end the conflict.

Writer's Options

1. Children often shed light on difficult situations. Write an essay from your own experience that recounts a time when a child taught adults a lesson.

2. Research some poetry written by children in concentration camps during World War II. What innocent wisdom do they display that is like the son's in this poem?

Tape or rent a few shows of *Sesame Street* (for one time only viewing). Analyze how this show attempts to teach cooperation and conflict resolution to young children.

Two Perspectives: Suggested Answers for Responding to Reading Questions

1. The degree of negative impact is a matter of opinion. However, if a girl were to consider this doll as her ideal woman, it is easy to see how plastic and shallow her goals would be. Boys are given toys such as GI Joe, Spiderman, and other comic book "superheroes" which, if adopted as images of real men, will leave the boys with unrealistic and possible negative images of maleness.

2. A rite of passage is an external action or public sign accepted by the community as a sign of maturity. Completing this passage also gives the individual new privileges and responsibilities.

3. Olds refers to the other boys as men because they are imitating their images of what men do.

4. Each poem shows that if a boy or girl does not shed the stereotypes of being a real man or a real woman, lives and happiness will be lost.

5. The poet's views are similar in that they show how children are corrupted by a distorted image of adulthood and how success and acceptance are gained in society. Piercy's poem shows the girl as a powerless victim. Olds poem shows the boys as active aggressors.

6. Answers here are open for discussion.

Using Specific Readings

"MARY CASSATT," MARY GORDON

For Openers

Every generation has had different vocations, occupations, and interests that have excluded one gender or the other. To begin the class discussion, brainstorm with the class about what areas today are either inaccessible or very difficult for either men or women to break into. Are there any areas left that don't at least pretend to be "equal opportunity" vocations? Do social pressures still exclude certain groups?

Teaching Strategies

Ironically, the language used to describe Cassatt's success is generally associated with masculine references. Indeed, the final sentence of the essay praises Cassatt as a "*master* of line and color." Discuss this with your students, and decide whether this phrase would be more effective if it were rewritten to be more gender-inclusive or whether it better represents her female success in a male world as Gordon wrote it. Also, can students think of synonyms for *master* in this context?

Collaborative Activity

Based on the information in this essay, and based on an analysis of Cassatt's paintings, ask students to write a more appropriate article that might have run in the *Philadelphia Ledger*, one that would replace the snippet in paragraph 1. Each group can read its article aloud to the class, and the class can then cut and paste these articles to create a full feature article.

Writer's Options

Gordon provides a number of reasons whey she does not want to believe that Cassatt and Degas were lovers. In your journal, write about a strong friendship you have had that was supportive and emotionally strong without being romantic.

Multimedia Resources

In order to help your students understand the points Gordon is raising about the technical aspects of Cassatt's work, bring in illustrations from art books, or any prints or slides your library might have. Contrast Cassatt's work with that of her contemporaries; does it seem more "feminine"?

Suggested Answers for Responding to Reading Questions

1. Gordon describes Cassatt as having the strength and insight that her male contemporaries could not have, simply by virtue of

being a woman; on the other hand, she had to fight a male-dominated system in order to be considered seriously. In other words, her femininity was both a blessing and a curse.

2. It is very doubtful that Cassatt would have received much recognition without her money and social standing, nor would she have had the opportunity to travel to Europe. The rest of the answers will vary.

3. Cassatt's story is "particularly female" because it depicts a male dominated world where accomplished females are reluctantly recognized. Gordon holds the opinion that her story is particularly American because Cassatt is emblematic of the indefatigable spirit found in other pioneer women.

Additional Questions for Responding to Reading

1. Are there female artists in other kinds of art (actors, singers, sculptors, novelists, dancers, photographers, poets, potters, composers, etc.) who have played the same role in their realm as Cassatt played for American women painters? Describe the impact of an artist you admire.

2. How does Cassatt's later life fit in with her earlier years in painting? Do you think she changed as she grew older, or can you identify traits that carried through her entire life?

"THE MEN WE CARRY IN OUR MINDS," SCOTT RUSSELL SANDERS

For Openers

Before assigning the essay, poll your students on the questions "These days, is it harder to be a man or a woman?" Have students brainstorm the advantages and disadvantages of each gender role. Record their findings on the board, and refer to them as the discussion progresses.

Teaching Strategy

In the first several paragraphs, Sanders writes about the betrayal of the men's bodies—how they give out, how they cannot function after a time. Talking about the physical nature of a person's body is not generally associated with male discourse; rather it is more often a characteristic of writing by women. Ask your class to find other places where Sanders uses phrases or techniques that might be considered "feminine." Why does he use them? Do they work?

Collaborative Activity

Ask students, working in groups, to discuss their male relatives' occupational roles. Do they see them as "warriors and toilers" or as something else? What other images can they come up with? In class discussion, identify or create other categories into which men might be classified. How much does social class determine the metaphors used?

Writer's Options

Write about your impressions of the power you have, and the power the opposite sex has over you. Use personal experiences to illustrate your points.

Multimedia Resources

Bring in other images of working-class men like those Sanders describes in his family. Possibilities include Billy Joel's "Allentown" video, the opening scenes from the movie *Rudy*, or scenes from *Roseanne*. Discuss whether the power Sanders describes is more a phenomenon of social class or of gender differences.

Suggested Answers for Responding to Reading Questions

1. Becoming either a "warrior" or "toiler" is the chief destiny of men. Both warriors and toilers take their orders from "bosses" — the people with the real power. This subordinate position helps Sanders see the relative powerlessness of women, since they, too, are under the leadership of those who make the decisions. Viewing his economic class in terms of power allows him to empathize more with women.

2. Answers will vary.

3. Sanders is an ally in that he also seeks an equal opportunity to challenge his ability to succeed in the marketplace. The women about whom he speaks would consider him an ally if they accepted the idea of equal rights applies to both genders. If they are only interested in the rights of their own gender, then they

will not allow Sanders to have an identity beyond the male stereotype.

Additional Questions for Responding to Reading

1. What does Sander's mean by "expansiveness" (p.380)? How important is that expansiveness to an individual and to a community?

2. How do Sander's sentence patterns and word choice create a sincere sounding voice?

"FIGHTING BACK," STANTON L. WORMLEY, JR.

For Openers

Ask students to discuss ways in which they have responded to physical threats. Discuss how men's and women's responses differ.

Teaching Strategies

1. Note that Wormley's argument depends on the ways in which he defines certain words and concepts, such as *fighting back, self-defense, retribution* and *justification*. Discuss the associations he makes with each of these expressions, and examine how they help organize his argument.

2. Show students how the essay approaches its thesis inductively — that is, how it raises questions in the introduction that are answered by the conclusion.

Collaborative Activity

Ask students to bring in clippings from newspapers and magazines that describe incidents of "fighting back" in self-defense. Have each group present one incident and ask the class to consider whether the response was justified in that case.

Writer's Options

Tell about a situation in which you found it necessary to fight back or defend yourself in some way (it does not have to be through physical confrontation). How did you feel? What did you learn? What would you do differently if this happened again?

Multimedia Resources

The movie *A Few Good Men* examines some of the issues Wormley raises. The film focuses on an incident involving some Marines disciplining another member of their platoon, and on the tensions between the male lawyer and female military supervisor assigned to the case. For the most part, the racial and gender lines are not clearly drawn, and students will be able to see the complexity and layers of tension between these groups in the military.

Suggested Answers for Responding to Reading Questions

1. Wormley may expect his readers to have the preconception that all African-American men are streetwise and know how to physically defend themselves. If he did not have this expectation he would not have opened the essay with the information that he came from "an upper-middle class black family" nor make the statement in paragraph 3, "I was brought up to *think*, not just react."

2. Answers will vary. Wormley says that what one loses by fighting back is not easily definable: compassion, tolerance, empathy. A woman who fights back may place more value on the less familiar feelings of assertiveness and empowerment.

3. As an African-American, Wormley does experience a certain kind of attack, one incited by racism. Morris chastises him for letting down his race. As an only child, however, Wormley might have been overprotected; he also would not have learned from siblings how to fight back. As an eighteen year-old, he might not have had much experience outside of his own neighborhood—and, more than likely, in his upper-middle-class neighborhood fighting was not a common sight.

Additional Questions for Responding to Reading

1. As an eighteen-year-old, did Wormley see himself primarily as an African-American, a man, or a member of the upper-middle-class? How about in the story he tells later, in paragraph 7?

2. What distinction does Wormley make between "fighting back" and self-defense?

3. What does Wormley mean by his "sadness" in the knowledge that he has learned to fight back? Do you find it sad? Explain your answer.

"Why I Want a Wife," Judy Brady

For Openers

Brainstorm to generate a list of family tasks and roles. Are they gender-related? Which are associated with men, and which with women? What is the origin of these associations? How might they be changed? Should they be changed?

Teaching Strategies

1. You will need to provide as much context as possible for this piece if you are to break through the student resistance it can prompt. (The Multimedia Resources suggestion will be helpful.) Set the stage for the reading by talking about the phases that the women's movement has gone through since the nineteenth century, and explain the tenor of the movement during the 1970s.

2. Since this essay frequently provokes strong student reactions, use it to talk about tone, irony, persuasion, and even satire. Lead into a discussion of audience based on an analysis of those who are offended by the essay and those who like it.

Collaborative Activity

Assign groups to the task of responding to Brady through the perspective of one of the following people:

1. The divorced man seeking a wife.
2. The husband.
3. One of the children.
4. A mother who works outside the home.
5. A single mother.
6. An older wife.

Present the responses to the rest of the class.

Writer's Options

1. Write your own spoof of Brady, explaining why you want a husband, a child, a grandparent, or some other person subject to stereotypes and disregard. Follow the structure that Brady uses: Discuss the person's routine tasks, then the "ideal" of their role, and finally the double standard inherent in their position.

2. Write a want-ad for a personal column in a newspaper. Describe in detail the duties and perks of the job that you are hoping to fill.

Multimedia Resources

In order to provide as much context as possible, get a copy of the Spring 1972 *Ms.* magazine (the issue in which Brady's article first appeared) from your library. Bring it into your class and discuss some of the other articles. What were the important issues that year? Discuss the advertisements; to whom are they appealing? To what qualities? How does Brady's essay fit in with the entire magazine?

Suggested Answers for Responding to Reading Questions

1. Answers will vary.

2. Although the essay seems bitter and outdated, far too many of the issues Brady raises are still around today. Some also say that there has been a backlash, returning women to the status that Brady deplores.

3. Answers will vary.

Additional Questions for Responding to Reading

1. What attitude toward children is revealed in this essay?

2. What is the purpose of the repetition of "I want a wife..."?

3. What evidence of Brady's socioeconomic background can be found in this essay? In what ways, if any, might the essay be different if somebody of a different class wrote it?

For Openers

Feminism is a term that has many meanings. Review the feminist movement with your class, showing how it began as a protest to a patriarchal society, then moved on to issues of equal pay for equal work and then to the unique experiences of growing up female and becoming a woman.

Teaching Strategies

1. In controversial situations, exploring the assumptions made in the argument is vital. Gutmann says, "The result has been a kind of feel-good feminization of boot camp culture, with the old (male) ethos of competition and survival giving at least partial way to a new (female) spirit of cooperation and esteem-building" (page 465). Earlier in the essay she writes about the "cultural problems" in the military. Discuss with students whether competition and survival are part of a military culture, a male culture, or something else. Likewise, does everyone agree that a spirit of cooperation and feeling good are part of women's culture?

2. If we agree that the purpose of the military is to engage in warfare to defend its country, can we expect and apply general social norms and a sense of equality to the military, or must it operate under a different code? Is it fair to apply a different gender-inclusive standard to the military?

Collaborative Activity

In groups, identify careers or organizations that were at one time exclusively or predominately male. Has the participation of women "dumbed down" these careers or organizations? In what ways has it changed or not changed them? For example, newscaster, sports reporters, road construction workers, etc.

Writer's Options

1. Ask students to write an essay in which they argue how if the sexes were segregated into separate military groups, both sexes could have more privileges and opportunities.

2. Gutmann says that the Internet has "become a haven for military samizdat about sex and other dicey matters." She quotes this on-line exchange, "we were never allowed to discuss our concerns openly because it would raise issues about the efficacy of mixing girls and boys and that was politically incorrect, a career-ending taboo." Write what you imagine the rest of such an on-line exchange might include. Take a position as to whether you think using the Internet this way is healthy or harmful.

3. Write a parody of how an all male or an all female organization should be transformed to include the opposite gender (sports teams, religious organizations, etc.).

4. As warfare becomes more technologically oriented and less hand-to-hand, argue why size and strength may no longer primary considerations among military personnel.

Multimedia Resources

1. Find an Internet site for recruiting in the military (http.//www.goarmy.com). Browse the different links. What are the ads like? What opportunities are promised for different genders?

2. Consider the movie, *GI Jane* (1997) with Demi Moore directed by Ridley Scott as a way our culture is depicting women in the military. Is this movie believable?

Suggested Answers for Responding to Reading Questions

1. According to Gutmann, the physical strength of women will never match that of men. Women will not be able to keep up in any training or performance situations. Because women become pregnant, they must take special jobs. Gutmann cites the Army newspaper, *Stars and Stripes* and Army personnel to support her claims. The situations she refers to seem quite plausible and realistic. However, since the source is an Army publication other less biased sources may be needed.

2. In paragraphs 3 and 4, the advantages of women in the military are raised. Gutmann does well to bring up this point of view early in her article so that she can have the chance to offset it with evidence against women in the military.

3. A compromise position should be raised and counterargued in order to complete the defense of her position.

Additional Questions for Responding to Reading

1. The physical differences between the sexes is cited as a major problem in including women in the military. Is this a strong enough argument against women in the military?

2. When Gutmann speaks of "re-evaluating" the training process of soldiers, she implies that the training becomes easier and that "feel-good" is substituted for "competition" is a negative situation. Do you agree that there is a loss if this shift has indeed occurred? Do you agree that women in military have caused this trend or that it is a paradigm shift that also exists in business and in education?

"MARKED WOMEN," DEBORAH TANNEN

For Openers

Go around the class and ask students to describe their own attire, stating whether they feel "marked" by their clothing or not. Generate a list of the different ways in which people are "marked" along with descriptions of what those marks mean. Keep a running tally to see how many men and how many women fall into the "unmarked " category. Is it really possible to be "unmarked"?

Teaching Strategies

Discuss with your students the organization and balance of Tannen's essay. First, she introduces her topic with personal observations; she then moves on to a quick explanation of the linguistic concept of "markedness"; finally, she plays off her own research in comparison with Fasold. Is the scholarly information handled appropriately for your students?

Collaborative Activity

The type of analysis Tannen undertakes in the first eight paragraphs lends itself to a group ethnology paper. Each group can choose a specific population on campus (or in the community), and gather data on the style of that group. Once they have gathered lists, impressions, and descriptions of between 10–15 members of their subgroup, they can analyze that style. What does it mean? Where does it come from? What sorts of contradictions arise from the ways in which this group presents itself?

Writer's Options

Have you ever been judged by your clothes? Have you ever judged another person by his or her style? Write about that experience, providing ample detail about each article of clothing, hairstyle, posture, and so on.

Multimedia Resources

Tape one of the many talk shows (*Oprah, Ricki Lake, Jenny Jones*, etc.), and analyze the appearance of the guests. Does the theme of the show relate at all to the attire of the guests? Do the differences between male and female attire that Tannen describes hold true on these talk shows?

Suggested Answers for Responding to Reading Questions

1. Tannen says that there is no standard dress, hair style, or make-up style for women. No matter what style of dress they choose, they are sending a signal to everyone of who they are. Men, on the other hand have a standard style that they can choose to wear. Answers will vary.

2. Answers will vary.

3. It is useful for Tannen to invoke a male and a scientific authority in order to position herself as reasonable, even moderate. In many respects, the findings Fasold reports are more feminist than Tannen's observations, making Tannen's argument not only more palatable, but more persuasive.

Additional Questions for Responding to Reading

1. What other types of words could be considered "marked"? Tannen gives examples of male words with female endings; can you think of other groups that are designated by a term that has a suffix?

2. Tannen raises the issue of maleness being considered generic, as representative of all members of society. What do you think of

this? Do you object to the term *he* when it is intended to refer to both men and women?

Focus: Is the Playing Field Level?

These three essays are written by women. Each essay emphasizes a different disadvantage to women in athletics. Is gender an issue in determining credibility in an argument about gender? Ask students to write their preconceptions on female athletes and male athletes before reading these essays. Discuss how preconceptions can affect one's reading of an essay.

"Title IX: It's Time to Live Up to the Letter of the Law," Donna Lopiano

For Openers

Ask students to list all the names of student athletes at your school. How do they know these students? Are they in class, in the dorm, or do they know them from local publicity? Decide if more male athletes are known than woman and why.

Teaching Strategies

1. The first four paragraphs not only establish the tone of the writer, but also give much background information on the issue. What kind of a personality does the writer seem to have? Discuss how the essay seems to be aimed at an audience that is not familiar with the situation or the legal history. Ask the class if this seems ironic considering the essay's place of publication.

2. What is the purpose of athletics at the college and university level? What does the student body have to gain by developing women's sports? Consider how equal investment in both men's and women's sports will affect your school as a community of learners.

3. Lopiano uses the verb "remember" throughout the essay. Discuss how this word is relevant to her purpose of not becoming too complacent with small changes.

Collaborative Activity

Ask students to go out on campus to interview "the person on the street" for their attitudes toward equal investment in student athletics. Return to class to share results. Are the results supported by Lopiano's essay?

Writer's Options

1. Interview a student athlete to find out how he or she chose your school. Learn what their practice and game schedules are like. Find out what perks they receive in terms of equipment, locker room area, services, and support. Do they feel they are treated well? After gathering information from this interview, write an essay in which you argue whether or not your school is "doing the right thing."

2. Write an essay in which you try to persuade your classmates that women's sports are as exciting as men's sports. Describe what they might see at a women's athletic competition. Keep in mind that your goal is to increase the number of people at the gate.

Multimedia Resources

Choose three universities or colleges. Access the home pages for these schools. How are the men's sports and the women's sports described? Which schools have made more progress toward equality? Are there equal numbers of men's and women's teams?

Suggested Answers for Responding to Reading Questions

1. Athletics are a source of scholarships and admissions to higher education. Participation in athletics can also contribute to academic success, improved self-esteem, better health, and leadership qualities. Schools are failing to comply with Title IX because their tradition is to support men's sports and because men's sports draw the crowds and alumni dollars.

2. Objectors to Title IX fear that men's sports will be cut. Lopiano suggests a stricter use of money and an elimination of wasteful practices.

3. Both Cahn and Lopiano agree that Title IX is being ignored. Both look to the need for more opportunities in women's sports

118

for health, educational, professional benefits. However, Cahn blames the press for the lack of respect and interest in women's sports. As a historian, Cahn's approach to this argument is based more on culture studies than is Lopiano's.

Additional Questions for Responding to Reading

1. Notice the question and answer structure of the essay. How is this pattern effective? Could this pattern be used in an academic paper?

2. Lopiano states that colleges need to spend more money on women's sports or they will be sued. Which is the stronger appeal, the one that appeals to financial security by avoiding "public embarrassment" and lawsuits or the one that calls for "strong ethical leadership"?

"THE NUMBERS DON'T ADD UP," MAUREEN E. MAHONEY

For Openers

Consider the need for the letter of the law and the spirit of the law. What is the value of each? How does the application of quotas relate to these values?

Teaching Strategies

Notice how clichéd terms help achieve an emotional response from the audience. Discuss the effects of such clichés as "ladies first,"(P3) "cut-and-dried formula," (P4), and "welcomed women with open arms" (P5). Discuss the effects of asking rhetorical questions.

Collaborative Activity

Investigate other campus organizations besides sports. How many men and women are there in each organization? Can numerical gender balance be achieved in these groups? What would be the advantages or disadvantages of such a balance?

Writer's Options

1. Answer Mahoney's question "And why do we think that it is so important to ensure that half the athletes are women when we don't seem to care that far less than half the dancers are men?" Use specific examples to support your opinion.

2. The concept of equality is at issue in *Cohen vs. Brown*. Establish a set of criteria to define equality. Then apply your criteria to this case.

Multimedia Resources

Review the news conference transcript of June 27, 1997 by E. Gordon Gee, president-elect of Brown University found at this address: (http://cgi-user.brown.edu/Administration/News_Bureau/1996-97/96-150t.html). Discuss the value of this transcript as evidence for an essay. It is important to practice the evaluation of Internet sources. Ask students to question whether this news conference is just a lot of boasting, or whether it reflects sound reasoning.

Suggested Answers for Responding to Reading Questions

1. Mahoney, writing for the popular press, does not assume familiarity with the case. She also takes the opportunity to review the main parts of the case in very biased terms so that she can predispose the audience toward her opinion.

2. Mahoney argues that equality is gained by more than just totaling numbers and looking for matching sums. Women could become victims to this kind of number game as well. Agreement with Mahoney will vary.

3. Lopiano and Cahn would probably argue that without strong and specific legislation, athletic directors will continue to take advantage of any money given to athletics and that the discrimination and competition won't stop unless forced to stop.

"YOU'VE COME A LONG WAY, MAYBE," SUSAN K. CAHN

For Openers

Formulate criteria by which to define a "real athlete." Are the criteria different for men and women? If the criteria are pretty much the same, why is there such a difference in attendance at men's and women's sports?

Teaching Strategies

This essay is about recent history as much as it is about sports. Point out the several historical references early in the essay that set the context for the discussion. How might a psychologist approach this issue? How might an economist? Show how different disciplines approach the same controversy by using different topics of conversation to gather evidence.

Collaborative Activities

1. Divide the class into sports such as baseball, soccer, gymnastics, track, and basketball. Have each group research how many girls and women's teams there are in your area. Use a *Yellow Pages* or a Sunday newspaper sports section as a source for this information. How do findings correspond to Cahn's conclusion?

2. Assign different popular magazine titles to different groups. Have each group review the issues of that magazine from the year 1972. How hot an issue does Title IX seem to be? Compare an issue of *Sports Illustrated* from 1972 to an issue from this year. What is the difference in articles or advertisements about women?

Writer's Options

1. Cahn says that "sports promised to reconnect women with their bodies and their power." Do you agree with this statement? How important is this connection? Do most men already have this connection?

2. Decide whether the issue of gender equality in sports is more an economic issue or a moral one. Use evidence from this article and from other sources for support. Decide how important athletic opportunities in our educational system are to our overall cultural well-being.

3. Respond to this statement: Educational institutions must hold to an ideal of social justice; therefore, they must lead the way and demonstrate fairness without outside legislation.

Multimedia Resources

The movie *A League of Their Own* reflects many of the issues raised in this essay. How do the women change as they continue to play? Consider how the team experience will affect a person's individual life. Consider how coaching women is different from coaching men. Does this movie break or reinforce stereotypes?

Suggested Answers for Responding to Reading Questions

1. The male model of sports is "marred by commercialism, corruption, and win-at-all-costs attitudes." Cahn states that the model is considered "sexist, elitist, and exploitative." Female leadership will emphasize and develop a relationship between women in sports and cultural development of gender relationships. Without fear of domination or suppression, women athletes can better develop personal goals and excellence.

2. Women's sports have more coverage in press articles and in advertisements. Title IX has brought the issue of inequality in financial support and facilities to the foreground. However, even if women's athletics were to have equal footing in high schools and colleges, opportunities to participate in professional athletics still remain lacking and lopsided.

3. Lopiano would agree with Cahn. She sees athletics as opening the gateway to education and careers for both men and women. Lopiano does claim that there are ways for college and university athletic programs to support both men's and women's sports without one jeopardizing the other, if administrators will respond to problems creatively. As more talented women are included in sport programs, the freer they will be to develop the ideal of the woman athlete.

Additional Questions for Responding to Reading

1. What is your response to Cahn's claim that the future of women's sports is a direct reflection of how women are being treated in general?

2. Does your own experience with women's sports concur with Cahn's claim that there is still a need for more opportunities for women in college athletics?

CHAPTER 6

THE AMERICAN DREAM

Setting Up the Unit: Building Ethical Appeals for Credibility

The readings that follow in this chapter describe and depict a wide range of individuals — and their attempts to discover how the American Dream fits into their lives. One common thread that runs through all of these readings is the necessity of explaining the narrator's position and relationship to the reader, to society, and to the American Dream itself. Each writer presents his or her own position, either implicitly or explicitly, and in the process makes readers more sensitive to his or her point of view. This process in which a writer communicates a sense of self to the reader in order to help support a point is called building *ethos*. It invites readers to be more empathetic, and it provides credibility for any stance that the writer may take later in the work.

The writers in this chapter use a variety of strategies to build their ethos for each piece. Some immediately state their affiliation with a certain ethnic, national, or other group. Others mention this kind of information only after they are well into the selection. Some are careful to consider their connections to their probable audience by including examples or appeals they know their audience (sometimes their adversary) will appreciate. Others explicitly try to encourage readers to identify with them in some way by focusing on similarities before moving on to differences. And, of course, many of the writers leave out information that might offend a reader who is not already predisposed to support their views. In short, each writer builds a picture of himself or herself that presents a reasonable, likable persona to readers. This is a valuable strategy for your students to use in all of their writing, particularly if they are trying to persuade readers with values different from their own.

As you and your students work through this chapter, ask them to notice and keep track of the ways in which the reader establishes his or her credibility with the audience. By the end of the chapter, they should have learned about a variety of strategies they can use themselves. Encourage them to do so by having them state somewhere in their papers (in a cover memo to the instructor?) who they picture as their reader, and the places where they feel they have built their ethos. Students could also try writing a paper as a letter to the person they have in mind, similar to the way in which the Declaration of Independence was framed.

Confronting the Issues

Option 1: Constructing Contexts

This chapter's readings present individuals who are prevented from achieving their versions of the American Dream because of race, class, gender, nationality, sexual orientation, age, personality, or some other specific factor. In many cases, the individual is also isolated from mainstream society because he or she does not share the social norms of a culture. William Graham Sumner (1840–1910), one of the founders of sociology in the United States, divided these norms into *folkways* and *mores*. Building on Sumner's ideas, twentieth-century sociologist Robert Merton divided norms into *positive norms* and *negative norms*. Discuss each of these concepts with your students, also adding definitions of *laws* and *values*. Encourage students to supply examples of each type of norm from their own lives. Then, as they read through the essays in this chapter, ask students to categorize the narrator's difficulties according to the kind of social norm he or she does not share with the larger culture.

- Folkways: Accepted standards of behavior passed down from one generation to another. These include familiar, routine, and sometimes trivial customs such as shaking hands as a greeting or making a comment when someone sneezes. Folkways vary, sometimes quite significantly, from culture to culture.

- Mores: Shared moral and ethical attitudes that establish the behavior necessary for maintaining social order. Examples of mores include coming to the aid of an accident victim, being honest in business dealings, and accepting the authority of the law.

- Positive norms: Things we should do, such as report a crime, and things we must do, like support our children until adulthood.
- Negative norms: Things we shouldn't do, such as cheat on an exam, and things we mustn't do, like commit murder.
- Laws: Norms that have been formally adopted as the rules that will govern society's behavior.
- Values: shared beliefs about what the society considers important and good. For example, individualism, democracy, the work ethic, achievement, success, and equality are all considered important values of American society.

Option 2: Community Involvement

Each of the readings in this chapter presents a different set of obstacles to achieving the American Dream. Break students into groups based on their interests in certain kinds of barriers, and have them research what can be done to alleviate at least one obstacle in your community or on your campus. Have the group agree on one (or two) barriers that it will tackle in the course of this unit, and ask them to devise a plan of action. For the rest of the unit, students can mobilize others to their cause. For example, they might use their writing skills to craft and circulate an applicable petition, they could provide educational services for those who may need them, or they may choose to carry a lobbying effort to influence local legislation. If the class cannot agree on a project, allow them to form several groups that support each other whenever possible.

Option 3: Cultural Critique

Ask students to collect and bring to class pictures of various icons and images of the American Dream: the flag, fireworks, apple pie, baseball, a two-car garage, fields of wheat, skyscrapers, and so on. Analyze each of these images to determine what and who are included, and what and who are excluded. What images do *not* reflect the American Dream? What colors are used? What are the main themes that tie all of these issues together? By contrast, what are the main themes that are implicitly excluded? Students should keep track of these lists, and add additional images as they encounter them in this chapter. What kind of stories can students construct based on the inclusion list, and based on the *exclusion* list? What statement does this make about the American Dream and who has access to it?

Option 4: Feature Film

Before reading any of the selections in this chapter, show the film *Roger & Me* to your class. It presents a view of the American Dream different from that offered in the readings. In it, documentary film-maker Michael Moore wants to depict the impact of the closing of a major GM plant in the town of Flint, Michigan. In doing so, he tracks down and interviews a wide range of people following (or just trying to find) their own version of the American Dream: beauty queens, unemployed factory workers, upper-level managers, country club golfers, spouses, jail wardens, and the soon-to-be-unemployed. He seeks, but never finds, the elusive Roger Smith, former CEO of General Motors. Ask your students to keep track of the ways in which the American Dream is characterized throughout the movie, and the way in which it manifests itself. Refer to this list, and build on it, as you work through the chapter.

Teaching "Two Perspectives"

In the two documents that follow, Thomas Jefferson and Martin Luther King, Jr., take on the social and political ills that plague each of their communities. Composed 187 years apart, these pieces of writing have helped shape our nation's history. Before reading them, ask students to write down brief impressions of these documents. What do they think these pieces are about? What kinds of issues do they think are covered in each? What has resulted from each? Compare the images students have of these texts with their reactions to the readings that follow. How close were their perceptions to the documents themselves?

"THE DECLARATION OF INDEPENDENCE,"
THOMAS JEFFERSON

For Openers

Discuss the statement "all men are created equal." Did the Founding Fathers mean to exclude women?

Teaching Strategy

You may use this document to introduce the concept of parallelism as a tool for adding stylistic emphasis. Have students identify parallel words, phrases, and clauses. Then, use their examples to illustrate how parallel language makes the Declaration's arguments more powerful.

Collaborative Activity

Ask students, working in groups, to read through the lists of grievances against the king, grouping the abuses into categories. Then, compare the groups' classification systems, fine-tuning them until there is agreement about which fall into each of these categories: economic, political, and human rights.

Writer's Options

Write your own declaration of independence. From what (or from whom) would you like to be free? Use a structure similar to that of the Declaration: First justify yourself philosophically, then list your grievances, and then explain your next course of action.

Multimedia Resources

Bring in copies for each student of the *Seneca Falls Resolution*, written by Elizabeth Cady Stanton and others in 1848. In it, the writers try to include women in the American Dream. Compare it to the Declaration of Independence. Are the structures similar? Are the arguments? The demands? Are the two documents equally persuasive?

"I HAVE A DREAM," MARTIN LUTHER KING, JR.

For Openers

Discuss in class why it is important to recall the early years of the civil rights movement. Ask students if they think any of King's dream has been realized. What parts still need work? Is there a different movement developing now? If so, what might it be?

Teaching Strategies

Discuss the following with your students:

1. Paragraph 3: Note the repetition of words and patterns. Note also that the paragraph is all one sentence. You might take this opportunity to explain the difference between a sentence that is simply long and one that is a run-on; a lesson in the use of the semicolon might be helpful.

2. Paragraphs 4–5: Point out the metaphor King employs here. Ask students why it is appropriate and effective.

3. Paragraphs 7–8: Might hostile listeners interpret the tone of this portion of the speech as threatening? How does King go on to address the possibility of such an interpretation?

4. Paragraph 14: Why does King distinguish between the treatment of African-Americans in the North and their treatment in the South? What, if any, are the differences?

5. Paragraph 35: Here King employs a rhetorical device called "distribution," a dividing of a whole into its parts. What purpose does this strategy serve here?

Collaborative Activity

Divide the class into groups, and assign portions of the speech (excluding the short paragraphs 26–34) to each group; instruct students to locate stylistic techniques such as figurative language, patterns of repetition, and allusions. Each group should identify the purpose behind King's choices and explain to the rest of the class how these techniques help contribute meaning to the work as a whole.

Writer's Options

As suggested in the headnote, the 1963 march on Washington, highlighted by King's speech, was a culminating moment in the civil rights movement and in many people's lives. Write about an event in your life that first made you aware of public affairs or incited you to some kind of public action.

Multimedia Resources

Students may need background about the early years of the civil rights movement (and King's role in it) as well as information about the August 28, 1963 march on Washington, during which King delivered this speech. If possible, show a videotape of the march, including the speech.

Two Perspectives: Suggested Answers for Responding to Reading Questions

1. The sheer repetition of the grievances, and their number create a powerful emotional appeal. Likewise, many of the word choices throughout rely on an emotional response; see, for example, the final paragraph. The document is also logical by presenting as its premise the principles of equality and then showing the breach of those principles.

2. The Declaration seems to be speaking for white men only, and if it were written today, the framers could be accused of racism and sexism. For their time, however, the framers expressed extremely liberal views.

3. Answers will vary.

4. Answers will vary.

5. Both of them address the person (or group) with whom they are battling, and both use the language of the dominant group; both work through existing channels to effect positive change, rather than working around the system; both balance reason and emotion in their appeals. They both share the dream of a nation where individual achievement and opportunity are possible for the sake of each citizen and for the prosperity of the nation.

6. Answers will vary.

Using Specific Readings

"WE MAY BE BROTHERS," CHIEF SEATTLE

For Openers

Have we responded in any way to the pleas included in this speech?
How? Which ones have we ignored? Which ones can we still do
something about?

Teaching Strategy

Discuss with the class the dominant imagery used throughout this
piece. Most of the metaphors rely on images of nature. Why is this
important?

Collaborative Activity

Divide the speech into parts, each covering a different aspect of the
ways in which the two cultures diverge. Ask each group to take one
part and articulate the main idea as well as the strategies used to
convey that idea. What is the overriding metaphor of that section?
What is the primary comparison and contrast the section makes? In
class, put the metaphors and the main points on the board,
demonstrating the continuity throughout the speech.

Writer's Options

This speech was recorded over a hundred years ago. If Chief Seattle
were alive today, what might he say? Write a new speech, using the
same structure as the old one.

Multimedia Resources

Find the section on Chief Seattle in the television documentary *500
Nations*. Show that part to your students and read the speech again in
class. How does the context provided by the documentary change
the way you read (or hear) the speech?

Suggested Answers for Responding to Reading Questions

1. The primary distinctions made in this speech are between the two groups' governing systems, religious attitudes, attitudes toward the dead, and the relationships with the environment.

2. The tone is predominantly resigned, with a reluctant acceptance of the fact that the white man is powerful, and is here to stay. It is also bitter about the ways in which the Native American ideals are subsumed into white culture. One dream it suggests for Seattle's people is that the white population might someday learn some of their lessons and listen to the ways in which Seattle's people interact with the surroundings; another is the hope that the white culture doesn't destroy all that the Native Americans value.

3. These sentences offer the smallest ray of hope: that the white men aren't really as bad as they appear, that they will learn to value the earth around them, that they won't continue on the path of destruction that they began.

Additional Questions for Responding to Reading

1. Which of the distinctions between cultures included in this speech are the most disturbing to you? Why?

2. What parts of the Native American culture described in this speech would you like to see play a greater role in American culture today? How might you work aspects of this culture into your own life?

"THE LIBRARY CARD," RICHARD WRIGHT

For Openers

Discuss the following questions with your students:

1. Are books pathways to other worlds? If so, is this sort of escapism positive or negative? Is the escapism that books offer different from the escapism offered by television and movies?

2. Can a person change on the inside without the outside world noticing the change?

Teaching Strategies

1. Isolation is a dominant theme in African-American literature. What might cause a person to become isolated from those around him or her? What causes Wright's increasing isolation from both blacks and whites — and from himself?

2. Ask students if they believe a class structure exists in the United States. If they do, ask how they think the various classes relate to one another. How does Wright's dilemma illustrate these relationships?

3. Ask students to consider changes in their own world views. What prompted these changes in their beliefs, values, and goals?

Collaborative Activities

1. Ask students, working in small groups, to analyze the key episodes in which Wright changes in some way. What prompts each transformation? Which changes are the most significant?

2. Wright uses dialogue in several scenes in this selection. Have students work in groups to create dialogue for some of the scenes that have none.

Writer's Options

1. In paragraph 81 Wright offers some insight into how he likes to learn. Write about the processes that you find most helpful in learning how to write. Which educational strategies work best for you?

2. Have you ever felt isolated from a group? Describe the circumstances, and explain your actions.

Multimedia Resources

To provide some historical context for your students, show scenes from *The Autobiography of Miss Jane Pittman*. How are her experiences like those Wright describes?

Suggested Answers for Responding to Reading Questions

1. Wright realized he has been restricted all his life. This realization creates a distance between him and others, both black and white. The changes are good insofar as they cause

him to grow intellectually; they are bad to the degree they make him more isolated. The American Dream is still very much out of his grasp, and now he understands this fact with increasing clarity.

2. Wright's "station in life" is that of a second-class citizen. He presents a passive, subservient, illiterate image to the whites around him in order to survive in a culture hostile to blacks

3. Wright's feelings of isolation make his decision to leave the United States seem inevitable. The rest of the response will vary. Those who believe in confronting injustice might suggest that he should have stayed; those more interested in Wright's personal happiness might applaud his decision to leave.

Additional Questions for Responding to Reading

1. How do books make "the world look different" for Wright? Is the difference good or bad? Or is it both? Explain your answer.

2. Wright's hunger for books is actually a deeper hunger for something else. What do you think he really wants? How do books help provide it?

3. In seeking to educate himself, Wright relies almost exclusively on books by white male authors. Do you think that by doing this he isolates himself further? Was this a good strategy?

"AMERICAN DREAMER," BHARATI MUKHERJEE

For Openers

Ask students whether they or any friends were ever teased about their names. How was their allegiance to family and family names compromised by the desire to belong?

Teaching Strategies

1. The organization of the essay is worth looking at. It begins with a personal narrative (1-14) and then discusses the difficulties in defining the concept of American identity (15-26); then, Mukherjee returns to her personal experiences by applying concepts of culture to her own situation (26-31) and then to the population at large. Discuss the importance of contextualizing specific situations within larger concepts.

2. A rejection of racial and ethnic groups of other racial and ethnic majority has insidious ramifications. Give examples from the text of how and why minority groups reject the majority, and how and why the majority considers themselves as superior and reject any other groups that are different. Discuss what is gained or lost in a culture when such a mentality has free reign.

Collaborative Activity

Identify differences among group members in terms of race, ethnicity, language, and religion. Then write a definition of "American" using your group as a model of your definition. Compare definitions with other groups. Do you think these definitions would be acceptable to the general public?

Writer's Options

1. Leaving home to go to college, students encounter different cultures and ways of thinking. Write a essay analyzing your own journey from your home community to the community of the university. What has your experience been like? What changes do you see in yourself that you embrace or resist?

2. Mukherjee says, "Rejecting hyphenation is my refusal to categorize the cultural landscape into a center and its peripheries; it is a demand that the American nation deliver the promises of its dream and its Constitution to all its citizens equally." Discuss the advantages and disadvantages of hyphenation.

Multimedia Resources

1. Multicultural issues involve art, politics, linguistics, and many other social and intellectual domains. Using the following Internet site, consider how this multicultural links Web page helps define multiculturalism as more than "the retention of cultural memory." The site is available at: http://www.albury.net.au/~curtinig/ethworld.htm# CULTURE.

2. Bruce Springsteen's song, "Born in the U.S.A." may provide a lively springboard for discussion or for a freewriting topic.

Suggested Answers for Responding to Reading Questions

1. Over the years Mukherjee says that she has been shaped by this crisis to become a stronger person and a better writer. She has resolved her crisis because she now sees herself as one who is being transformed and helping to transform her new culture. Native-born Americans too face changes in culture. For example, when they marry, move to different parts of the country, or go away to college. There are many ways in which we reconstruct ourselves as we grow.

2. Mukherjee has lost the traditions and rituals of her family and the "visceral" connection that bound her to it. By leaving, she lost the role that was established for her by her family. She no longer has the certainty of what she will become. She speaks primarily for herself, but in this particular story, one can extrapolate others' immigration stories. Her story is very particular to her own experiences, but since she discusses them in broader terms as well, one can draw more general conclusions about the issue of enculturation.

3. Responses will vary.

Additional Questions for Responding to Reading

1. Mukherjee relates the various ways in which community experiences have shaped her. List the different groups with whom she has lived and write a short description of the impact each community had on her. Are there commonalties among these groups or did each provide a distinct experience?

2. Consider the various communities of which you are a part. How do different communities shape your self-identity?

"THE MYTH OF THE LATIN WOMAN: I JUST MET A GIRL NAMED MARIA," JUDITH ORTIZ COFER

For Openers

Begin the class with a freewriting period, asking the students to recall a time they were treated as an "other." Ask them to describe the circumstances, the dialogue, and the feelings evoked.

Remind the students that Puerto Rico is a part of the United States and that Puerto Ricans are U.S. citizens.

Teaching Strategies

1. Cofer's resentment for the way she has been treated is communicated often through a sarcastic tone. For example, she writes, "I don't wear my diplomas around my neck for all to see"(11). Find other places where this sarcasm emerges. Is it effective? Consider how humor in her opening paragraph and sarcasm in other parts of the essay reveal her strong feelings without alienating her audience.

2. The essay could easily become fragmented into separate anecdotes, but transitional devices hold it together. Comb through the essay to highlight transitional devices within paragraphs and between them.

Writer's Options

1. Write your own version of this essay, showing the myths about a group to which you belong – for example, student athletes, sorority or fraternity members, single parents, or your own ethnic group.

2. Cofer writes, "My goal is to try to replace the old stereotypes with a much more interesting set of realities. Every time I give a reading, I hope the stories I tell, the dreams and fears I examine in my work, can achieve some universal truth that will get my audience past the particulars of my skin color, my accent , or my clothes"(13). She says she seeks to open "some avenue for communication." In light of her purpose for writing and publishing, argue the need for an open literary canon and multicultural readings in all disciplines. What effect is a multicultural perspective in your own college education going to have on you beyond graduation?

Collaborative Activities

The essay is not very complicated in the way it unfolds: Cofer identifies many myths and describes each one. The essay's organization is straightforward. To appreciate what kind of prewriting may have been used to write this essay, ask groups to "unwrite it" using inventions strategies of mapping, outlining, dramatizing, and listing. Supply transparencies to each group so that they can show their results on an overhead projector.

Mapping: Draw a circle in the center of a page and write in that circle "Latina myths." Draw circles around the center circle, labeling each with one of the myths. Any illustrations are illustrated in boxes with a brief phrase to identify them, such as "Daddy sings."
Outlining: Use a conventional outline format to parse out the sections of the essay.
Dramatizing: Using the text, answer the questions: Who? What? Where? When? How? and Why?
Listing: List the myths and the corresponding illustrations under each list item. Listing is an informal version of an outline.

After each group has presented, ask student to write a reflection on which invention strategy most appeals to their own writing process and why.

Multimedia Resources

Ask students to view *West Side Story* so that they can become familiar with the references made in this essay. Since this film is considered an adaptation of Shakespeare's *Romeo and Juliet*, discuss the idea that from a particular story a universal truth can emerge.

Suggested Answers for Responding to Reading Questions

1. Movies, songs, and jobs held by unskilled or educated Latinas, and cultural traditions have perpetuated the myth that Latinas are sex objects who cook and clean. The danger of objectification is real, as one loses personal identity within one's community.

2. Cofer is claiming that one cannot blend, fade, or melt into the dominant culture, but that one is always associated with the place of origin. This may also be true for other groups, but it is a matter for discussion and shared opinions.

3. The fact that Cofer feels compelled to write this essay suggests she still feels some "cultural schizophrenia." The very first paragraph shows how she is torn between cultures: while she is offended as a Latina, she "managed my version of an English smile."

1. Cofer states that she hopes that by telling her stories "the dreams and fears I examine in my work can achieve some universal truth that will get my audience past the particulars of my skin color, my accent, or my clothes." Has her essay achieved this goal for you as a reader? What universal truth does her essay suggest?

2. To what extent do you agree that the problem of prejudice toward Spanish-Americans is particularly severe for Latina women?

"LIMITED SEATING ON BROADWAY," JOHN HOCKENBERRY

For Openers

Ask students what they think of Hockenberry's reactions. Do they think he should have acted differently? (Keep in mind that he eventually sued the theatre and won a settlement that required increased access for disabled patrons.)

Teaching Strategies

1. In paragraph 11 Hockenberry makes a parenthetical comment regarding the art community's relationship to defecation, which you may need to explain to your students. Some artists, particularly performance artists, have built their careers and their reputations by pushing the limits of what is socially acceptable; one in particular was granted funding by the NEA even in the face of public outcry because she smeared herself with her own feces on stage. In paragraph 13 Hockenberry gives his opinion of certain types of performance artists, people who carry "around bottles of urine just to make a point about society."

2. What do your students think of the idea of a blind person climbing Mount McKinley, or people with various disabilities taking dangerous voyages alone, or competing in athletic contests? What motivates these people? Should they be encouraged to compete, to explore their own possibilities, or should they learn to accept their limitations?

In paragraph 9 Hockenberry describes himself as he thinks the ushers might remember him. Ask students, working in groups, to take the perspective of someone who might have witnessed the altercation at the theatre, and present his or her version of the story. What about the friend? What about the other theatre patrons? What about the person in the ticket office?

Writer's Options

Ask students to write about a time they felt left out because of some physical limitation. What was the nature of the limitation? From what were they excluded? Was there anything they could do about it?

Multimedia Resources

1. Bring in the text of the Americans with Disabilities Act. Ask students to read it before reading the Hockenberry article. Does the legislation make sense? Does it seem reasonable? Ask students to revisit their responses after reading Hockenberry's experience. Did their attitudes change at all? Is there still anybody who might be excluded?

2. What do your students think of the idea of a blind person climbing Mount McKinley, or people with various disabilities taking dangerous voyages alone, or competing in athletic contests? What motivates these people? Should they be encouraged to compete, to explore their own possibilities, or should they learn to accept their limitations?

Suggested Answers for Responding to Reading Questions

1. Hockenberry is most angry at the hypocrisy of the arts community, which he characterizes as people who make their living out of difference and controversy, without making the connection to real, nonfictional people with their very physical and other problems. Hockenberry would expect members of this community to be more sensitive than others, because they *should* know better.

2. Hockenberry would like access to more functions and places; he would also like to be treated with more respect. The situation

at the theatre could have been averted if the person who sold him the tickets had asked more appropriate questions; he still would have been denied access, but he wouldn't have been so humiliated by the process. The rest of the response will vary.

3.   Answers will vary.

Additional Questions for Responding to Reading

1.   Hockenberry uses a counterexample from a puppet show in East Jerusalem (7). What does he accomplish by using an illustration from another country, another culture? Would his point have been made more effective if Hockenberry had used an example from the United States?

2.   This essay was published in *The New York Times.* Do its readers seem to be an appropriate audience? Why, or why not? If not, what readers might be more appropriate?

"THRIVING AS AN OUTSIDER, EVEN AS AN OUTCAST, IN SMALLTOWN AMERICA," LOUIE CREW

For Openers

Crew writes that it "is easy to confuse sensible nonviolence with cowardly "(15).   Ask your students to explain what these two types of behaviors are and to articulate the difference between them. They can use examples from their own lives to illustrate the distinction.

Teaching Strategy

Some students may resist talking and writing about issues of homosexuality. If this is the case in your class, focus instead on how the essay is written and what types of appeals Crew uses. Who is his intended audience? How has he managed to reach so many people who were not open to what he had to say? Often students cite the physical nature of the homosexual relationship as the most disturbing part of the issue to them; point out that Crew mentions love and commitment, but not sex.

Collaborative Activity

Beginning in paragraph 16, Crew outlines several of his survival and educational tactics. Divide the class into several groups, and assign each group to a different part of the essay. Each group should then analyze the strategies used by Crew and his companion to help reeducate their new community. What is the strategy? How might students use a similar tactic to gain more voice, more power, more respectability? Discuss the possibilities as a class.

Writer's Options

1. Crew discusses the trade-offs he has made throughout the essay, but particularly in paragraph 34, where he mentions their relative lack of privacy. Write about the kinds of trade-offs you are willing to make in order to further a cause you believe in. What are the causes worth fighting for? What have you done? What stands have you taken? What personal sacrifice did it involve?

2. How did you respond to Crew's use of the word *spouse*?

Multimedia Resources

Clearly, Crew's religious affiliation is important to him, and he says he derives much of his self-worth from this devout belief. Bring to class either the printed version of *Ecclesiastes* (mentioned in paragraph 35) or a recording of the Byrds or Pete Seeger singing *Turn, Turn, Turn*, and ask students to identify the ways in which Crew lives and acts within the precepts of that verse. Where does he show his commitment in this essay?

Suggested Answers for Responding to Reading Questions

1. When Crew and his companion first moved to Fort Valley, they were alternately taunted and dismissed. It was not until a year after they began to fight back that they won the grudging acceptance of community members. It might have been easier for them to live in a large city, but people can rarely pick their ideal place to live when they are constrained by their jobs; likewise, Crew's sense of civic duty might not have manifested itself in the same way if he had had a larger gay and lesbian community to rely on.

2. Answers will vary.

3. Crew and his companion started surviving the day they began to fight back; they didn't let the ignorance and hatred of the townspeople keep them locked away. When others sought them out, allowed the couple to reeducate them, accepted the larger

gay community rather than just the two "clever queers," they began to thrive.

Additional Questions for Responding to Reading

1. Crew writes in paragraph 17 that he believes "that only organized and sustained resistance offers much hope for long-range change in any community...[t]he random act is too soon forgotten or too easily romanticized." Explain what he means by this, using examples.

2. In paragraph 22 Crew says that he and his companion feel as if they have to be "twice as effective as [their] competitors just to remain employed at all." Do you agree with this? Why or why not? Is it true for other groups — for example, African-Americans or women?

3. In what ways do Crew and his companion conform to the stereotypes associated with gay men?

4. How do you think Crew would describe the American Dream? How close is he to achieving it?

"HOW IT FEELS TO BE COLORED ME," ZORA NEALE HURSTON

For Openers

Ask students to describe their reactions to the title of this essay. Does it make sense grammatically? What connotations does the word *colored* have?

Teaching Strategies

Discuss the following with your students:

1. Explain the point Hurston is making by differentiating herself from other African-Americans in the first paragraph.

2. Why does she say in paragraph 8 that the position of the whites is more difficult?

3. What are the circumstances under which Hurston feels most acutely colored?

4. What is the "cosmic Zora" (14)? How else might you phrase this concept?

Collaborative Activity

Ask each group of students to figure out all the different ways in which Hurston uses the word *colored*. What does it represent at different points in the essay? What synonyms could be substituted in different places? As a class, discuss the different interpretations of the word, showing the numerous meanings it holds.

Writer's Options

In paragraph 17 Hurston uses the metaphor of brown bags. She then goes on to list the items that "the Great Stuffer of Bags" has metaphorically placed there. Ask students to generate their own lists, creating items that represent parts of themselves, parts of their memories, and parts of their dreams. They can use this list in their next paper, if appropriate.

Multimedia Resources

A 1985 play by George C. Wolfe called *The Colored Museum* consists of a series of one-act "exhibits" that satirize stereotypes of African-Americans. A number of its vignettes poke fun at the different meanings of *coloredness*. Pick the scenes that you think your students would respond well to, and show (or read) them to the class. How do they support or contradict Hurston's argument?

Suggested Answers for Responding to Reading Questions

1. Being colored often just means being different; it holds as many positive implications as negative ones for Hurston. As she matures, she becomes aware of some racial differences in attitudes and preferences, primarily cultural ones, but she prefers to identify herself with an all-encompassing race.
2. She does not view her "differentness" as a detriment to achievement; rather, she accepts white society's attitudes about her race as a challenge.
3. Hurston suggests there are times when race is irrelevant.

Additional Questions for Responding to Reading

1. Exactly when did you become aware of your own racial or ethnic identity? How?

2. What message is Hurston sending to other members of her race? What flawed attitudes does she perceive among them? How does she propose to eliminate these attitudes?

3. How does Hurston respond when she is confronted by discrimination? Do you agree with her actions?

Focus: Do We Need Affirmative Action?

Ask students to write a list of questions they have about Affirmative Action. Explain the legal origins of it. In groups, see if answers can be found for some student questions. As students read the following essays, ask them to seek within these essays ethical appeals that could support an answer to their remaining questions.

"SINS OF ADMISSION," DINESH D'SOUZA

For Openers

1. Explain the play on words in the title. The usual reference to erring is "sins of omission or commission."

2. Introduce D'Souza as an author of popular books about higher education and the literary canon. He argues well from a conservative perspective and is very critical of administrators and faculty members who, he thinks, hold students hostage to radical social beliefs.

Teaching Strategy

D'Souza's essay seeks to persuade its audience of the existence, scope, and implications of the problems with Affirmative Action. Study assumptions made in his argument and discuss whether they are acceptable. For example, in paragraph one, he refers to Affirmative Action as, "the issue that is central to understanding racial tensions on campus and the furor over politically correct speech and the curriculum." Are there other causes for racial tension? Does politically correct speech find its source in Affirmative Action?

Collaborative Activity

The use of impassioned words and words that carry negative connotations is a persuasive tactic in this essay. Divide the essay's paragraphs among the groups and ask students to find such words and phrases and replace them with more neutral terms. What happens? Discuss the value of slanted word choice and the responsibility of both writer and reader to be sensitive to its use. You can start out with some examples, such as "lofty goals"(3), "mediocre credentials" (4), "cruel irony"(6).

Writer's Options

1. Respond to D'Souza's statement that ethnic diversity is sought by colleges to "prepare young people to live in an increasingly multi-racial and multicultural society." Is this preparation something you consciously sought when you picked your university? How is it important?

2. If you are not a traditional "young person" attending college, how do you respond to D'Souza's statement?

3. Affirmative Action is discussed primarily as a responsibility of college administrations and admissions officers. Write an essay explaining the responsibility and possible responses of students and faculty to multicultural issues.

Multimedia Resources

Ask students to search links provided by *Atlantic Unbound* by keying in the subject *affirmative action* in a search engine of their choice. The *Atlantic Unbound* site will provide the history and current dialogue on Affirmative Action. The address is:
http://www.theatlantic.com/atlantic/caverj.html/

Suggested Answers for Responding to Reading Questions

1. D'Souza states that in order to achieve "proportional representation," colleges must lower their entrance requirements, leading to "misplacement " of students, leading to higher drop out rates among minorities. Minorities may form groups to help each other through college, but this leads to "separatism and self-segregation." D'Souza does not take into account that SAT scores may not be a good measure of student abilities and pre-college experience. Isolation may be caused by bigotry on college campus. Answers will vary in discussion.

2. Answers will vary.

3. Steele may very well agree with some of D'Souza's points because Steele feels that Affirmative Action has negative effects on those it seeks to help.

Additional Questions for Responding to Reading

1. What might D'Souza mean by "resurgent bigotry"? How would this kind of prejudice show itself on your campus?
2. What advantages or disadvantages are associated with ethnicity and with socioeconomic status? Which do you think is the most influential in contributing to a person's success or failure in college?

"A NEGATIVE VOTE ON AFFIRMATIVE ACTION," SHELBY STEELE

For Openers

Looking at the word choice and overall tone of the essay, discuss how readers might characterize Steele. His calm tone, sophisticated vocabulary, and organized discussion create confidence and credibility in the audience. Would most of your students agree?

Teaching Strategy

Refer to Martin Luther King's "I Have a Dream" speech (p. 431), focusing in particular on two passages: "We can never be satisfied as long as our children are stripped of their selfhood and robbed of their dignity by signs stating 'for whites only'"(14) and "I have a dream that my four little children will one day live in a nation where they will not be judged by the color of their skin but by content of their character"(20). Discuss how Steele's position correlates with, complements, and/or conflicts with King's speech.

Collaborative Activity

It is important to distinguish the specific terms that Steele uses. Divide the following terms among student groups. Ask students to explain the meanings of each word, basing the meaning on the essay's context and on dictionary or popular definitions: *moral symmetry, real diversity, proportionate representation, racial representation, racial development, victimization, parity, equal opportunity.*

Writer's Options

Describe the kind of learning environment and community you hope your children will have available to them. What action do we need to take now to move in the direction you are proposing?

Multimedia Resources

Using the Internet, go to the Web pages for the *Center for Equal Opportunity* (www.ceousa.org) and the *Center for New Black Leadership* (www.cnbl.org). Browse through their various links. What are the issues for these two groups? How would they respond to Steele's views? What other debates arise from the topic of Affirmative Action?

Suggested Answers for Responding to Reading Questions

1. Steele argues that Affirmative Action creates new problems and does not solve the deepest problems of discrimination. Affirmative Action is seen by many as reverse discrimination, which causes resentment rather than good will. It implies inferiority of black college applicants. Preferential treatment raises doubts in the students themselves about their qualifications and chances at success. As an apparent solution to education of black students, it does not solve the need for "racial development" over "racial representation." Steele believes that equal opportunity at all levels will lead to the development of all students and empower students who might otherwise continue to see themselves as victims.

2. D'Souza's essay claims that unqualified minority students who are admitted under Affirmative Action struggle to "keep pace" with other students. D'Souza's essay supports Steele's essay with its references to the anti-Asian policy at the University of California and also with its reference to surveys and interviews published in *The Chronicles of Higher Education*. Lawrence and Matsuda's essay refers to the Truth Commission which confronted past wrongs in South Africa. Their argument that a country must admit to its collective guilt undercuts Steele's essay.

3. This same generalization does exist in the other essays. Lawrence and Matsuda argue against the resistance to admit to the need for reparation. They talk about the desirable result of "spiritual wholeness of all citizens"(p.485), but do not discuss

147

resentment and dissention on a community. D'Souza argues the lofty goals of Affirmative Action are impossible to achieve in light of the possible damaging effects he lists. Steele does not seek an either/or resolution, but rather points out the need for new solutions and social policies. Either/or resolutions are often oversimplifications of most problems and their solutions.

Additional Questions for Responding to Reading

1. What is meant by a "Faustian bargain"? How does this reference intensify the frustration that Steele wants to convey? How does this reference fit in with other references and the overall tone of this essay?

2. Steele states, "The impulse to discriminate is subtle" (22). What is your experience with subtle discrimination? What strategies have helped stop or change it?

"THE TELLTALE HEART: APOLOGY, REPARATION, AND REDRESS," CHARLES R. LAWRENCE III AND MARI MATSUDA

For Openers

Discuss the ethics involved in reparation for wrongful deeds.

Teaching Strategies

1. The essay argues by using analogy. Identify the analogies and discuss if they are acceptable parallels to the Affirmative Action issue. Some analogies for discussion are *The Tell-Tale Heart*, pollution disasters, cases of fraud, and child abuse.

2. The essay uses rebuttal effectively. Point out how the objections to Affirmative Action are listed in paragraph 23, then addressed individually from paragraph 30 on. After a close reading of these sections, discuss the effectiveness of this rebuttal strategy.

Collaborative Activity

Compare experiences with what one might call "rituals of civility." What established practices have you learned from different communities on how to keep the peace and well-being of that community—or, as the essay puts it, "the emotional and spiritual balance of healthy souls?" How might those rituals find their counterparts in the Affirmative Action issue?

Writer's Options

Write an extended analogy to demonstrate your position on Affirmative Action; then explain the analogy and claim your position.

Multimedia Resources

The movie, *The Power of One*, depicts how one grandfather and one boy defy an unjust system and personally fight for justice in South Africa. View this movie as a class, stopping between scenes to give students opportunities for freewriting and reflection.

Suggested Answers for Responding to Reading Questions

1. Steele would probably not be swayed by Lawrence and Matsuda's argument. Steele holds that this argument does not take into consideration enough of the results of Affirmative Action. Other answers will vary.

2. Answers will vary. Some students may refer to the Rodney King beatings, police brutality during the peace marches of the sixties, harassment of black youths, and the Denny's Restaurant discrimination case.

3. Lawrence and Matsuda would probably welcome such practices because in their view these gestures heighten the society's sense of history and fulfill present-day obligations to the past. Other authors in this section do not see responding to historical events as a solution to today's problems. They find that such apologies and paid reparations contribute to the problem of discrimination rather than resolving it.

Additional Questions for Responding to Reading

1. Without reading the introduction about the authors, could you tell they are lawyers? How do they establish credibility with the reader?

2. Lawrence and Matsuda state, "There is a wisdom beyond human comprehension that determines what pain to exact for human transgression. Few cultures exist that do not have some notion of judgment. Actions have consequences. This is a law

of physics we transmute to a law of our lives." Is this true in most human relationships?

CHAPTER 7

THE WAY WE LIVE NOW

Setting Up the Unit: Adding Details

from Popular Culture to Provide Context

This chapter offers a variety of perspectives on the way we live now, and many of them seem to comment directly on one another. One common thread that runs through almost all of the readings is the stylistic device of piling up details to provide a context for the writer's perspective. For example, Jenny Lyn Bader illustrates what she means by heroes by describing them through a catalog of objects from her childhood: toys, books, movies. Arthur Levine lists the events and products by which younger generations define themselves; Pico Iyer gives us a vision of a merchandise-driven global village; and so on throughout the chapter.

One strategy for setting a stage is using product names that strongly identify a certain time, feeling, class association, or age group. Even something as simple as a beverage choice can say a lot about a writer: Does she drink Cherry Coke, Pabst Blue Ribbon, Evian, Snapple Iced Tea, Grape Nehi, a vanilla malted, Diet Dr. Pepper, a lime phosphate, Mountain Dew, or white zinfandel? Likewise, describing certain clothing choices can help the reader picture the kind of person who is speaking or being spoken about. Calvin Kleins or Levi's? Gap or Big Dog? A sweatshirt with the logo of the Michigan Wolverines or the San Jose Sharks? Students are generally pretty sensitive to the differences in these markers, so they should be encouraged to try using specific product names to strengthen their writing. Used appropriately, brand names can help them achieve a level of specificity their papers sometimes lack otherwise.

Another useful technique is the one Levine uses to frame his entire argument: using a familiar event to locate the subject of the writing. Students can use a current event as a reference point for their writing, exploring the connections between something that is happening in their lives and something that is making national news. Movies, songs, magazines, and television shows can also suggest connections among students' lives, the way they see themselves, and the issues they are writing about. These connections can be used to illustrate similarities or differences among the different generations.

Of course, over-reliance on examples from popular culture can weaken, even trivialize, student papers. You will need to caution your students against this. Nonetheless, it is worthwhile to have your students experiment with mentioning specific products to characterize a particular person, group, place, or time period. This strategy can help them focus on providing useful details for their readers and it may also help them discover what is unusual in products they see every day and examine their own relationship with these products.

Confronting the Issues

Option 1: Constructing Contexts

In journalistic terms, a story's newsworthiness depends in part on the question of proximity—that is, on how geographically close its events are to readers. War, famine, natural disasters (earthquakes, floods, and so on), and terrorist attacks are always seen as important in the abstract, but when these events occur close to home (to people near us—and, therefore, to people like us), their impact is much greater.

Thus, readers will feel shock and compassion at news stories about the fighting in Bosnia, a poison gas attack in the subways of Japan, and a devastating earthquake in the former Soviet Union, but when asked to rank social problems in order of importance, most readers will pick something close to home. For instance, urban readers will probably rank a story about street crime or drugs the highest, while rural readers might cite features on water pollution, droughts, or farm foreclosures. After explaining this concept of proximity to your students, and extending it to encompass emotional as well as geographical closeness, ask your students to rank in order of importance the following issues, each of which represents an ongoing problem in the United States today.

Afterwards, tally your students' votes to determine which issues they see as most—and least—pressing. Why? Which do they believe threaten U.S. society? Which threaten their own communities? Their families? Ask students to articulate what determines the importance of an issue for them. How optimistic are they about the chances that any of the problems they rank as most pressing will be solved? Which, if any, do they expect to disappear within ten years?

- Racial discrimination
- Teenage suicide
- Crack use
- Domestic violence
- Alcoholism
- Discrimination against gay men and lesbians
- Acquaintance rape
- Homelessness
- Terrorist attacks on abortion clinics and providers
- Corrupt municipal government
- Overcrowded jails
- Environmental decay
- Discrimination against women in the workplace
- Welfare reform
- Teenage pregnancy
- The right to die
- Inequities in the public education system
- Police brutality
- Obstacles facing persons with disabilities
- Violent crime
- Access to health care
- Bilingual education
- AIDS
- Unemployment

Option 2: Community Involvement

Many of the problems discussed in this chapter arise from a misunderstanding or lack of communication between generations. It has been said that each generation has a lot to teach the others, if only they will listen. For this project, ask students to find out what organizations exist in the school or in the community for bringing the generations together. For example, do churches or fraternities or sororities sponsor visits to nursing homes or hospitals? Are there chess or card clubs that specifically reach out to a mixed-aged audience? After students have located several places where the generations interact, ask them to divide themselves into groups, each one covering a different type of organization. They can then write brief ethnographic reports of what happens in these meetings, first interviewing the participants to learn what each hopes to get from their cross-generational encounter. (Students can even become involved in the programs they are studying.)

Option 3: Cultural Critique

Ask students to bring in a visual image that represents for them the most serious problems they feel they face in their lives. They can create the images themselves, or they can bring them in from advertisements, cartoons, album covers, posters, cereal boxes, newspapers, or anything else with a graphic image. Put all of the images up at the front of the room, and look for similarities. What trends do your students see within these images? What impulse do many of them seem to share? Are there recurring themes? Are there any that don't seem to fit with the rest? What qualities set those apart? After the class has had a chance to look at the images without any explanation of what they stand for, go around the room and ask students to describe the issue they felt their image represented. Keep a separate list of the issues, and see how they match up with the images (Although a variety of images will be represented, a few problematic issues — money and violence, for example — will probably be dominant.)

Option 4: Feature Film

Although it is long, the film/documentary *Hoop Dreams* articulates many of the issues that run through the readings in this chapter. The film focuses on two boys in the Chicago area who are tapped at very early ages for their talent for playing basketball. Both are offered scholarships to a Chicago-area private high school, but only one can stay in the school. The film traces the boys through high school and part of college, showing their heroes, their dreams, their nightmares, their economic difficulties, their problems at home, and their successes and failures. Students can compare the stories of these two young men with the images we see of basketball stars like Michael Jordan, Charles Barkley, Dennis Rodman, and Shaquille O'Neal. They can also trace similar struggles through the selections in this chapter.

Teaching "Two Perspectives"

Before reading "Larger than Life" and "The Making of a Generation," ask students to respond to each of the student voices at the beginning of the chapter. Where do they see themselves fitting in? Do any of the student quotations make them angry? Do any speak for them? How would they respond to these voices if they had the chance? The two readings that open the chapter also discuss ways in which generations view themselves, their definitions of themselves, their heroes, and their predecessors. Ask your students to write down their personal definition of their own generation before reading Bader and Levine.

"LARGER THAN LIFE," JENNY LYN BADER

For Openers

Much of what Bader describes as the current generation's relationship to heroes has been played out in the O.J. Simpson trial. Discuss with your class those aspects of O.J.'s hero status, and his fall from grace, that Bader anticipates and projects. It might be important to point out that she wrote this piece before the time of Nicole Brown Simpson's and Ronald Goldman's deaths. How, then, does the O.J. story fit with Bader's argument? How does it contradict her?

Teaching Strategies

One of Bader's writing techniques is to combine disparate elements within a sentence. Point out the following examples to your students, ask them to discuss the effect of these juxtapositions, and ask them to locate additional examples:

> Paragraph 2: "...the world was freshly populated by gadgetry and myth."
>
> Paragraph 17: "Of all the myths I happily ate for breakfast..."
>
> Paragraph 22: "Samuel Adams became a beer, John Hancock became a building..."
>
> Paragraph 46: "Hercules can't go on "Nightline" and pledge to stop taking steroids."

Collaborative Activity

As a class, create a taxonomy of the different categories that Bader uses: hero, icon, idol, role model, and so on. Divide the class into groups, and ask each group to compose a working definition of one of these "types," a descriptive list of its qualities, and a list of the people (or nonpeople) who might fit into this category. Pool this information with that from the rest of the class, refining and clarifying where necessary. Students can use these distinctions and examples in their next paper, if applicable.

Writer's Options

In paragraph 13 Bader's friend Katrin asks if people still have heroes. Respond to her in your own words. Do you have heroes? Do you think other people do? Do you think we need them?

Multimedia Resources

Bring in a variety of children's books (some current, some classic) intended to introduce heroes to young children. How are the images, illustrations, phrases, and stories—and the heroes themselves— different in current and classic books? Which do students see as generally "truer" or more valuable educationally, emotionally, or socially? What are the strengths and weaknesses of each?

"THE MAKING OF A GENERATION," ARTHUR LEVINE

For Openers

Discuss the opening sentence with your students. Do they believe that each generation is defined by social events? Or is there something else that defines a generation? What might it be? Where, and how, do social events play a role?

Teaching Strategy

You might need to provide some additional information for students who aren't fully aware of each of the events that Levine describes. For instance, if you have a sizable population of international students, or if you have students who have limited access to television, those students might not readily understand the importance that Levine and his subjects ascribe to each event. Take a quick survey of your students to see if they require more background than this article provides.

Collaborative Activity

Divide the class into five groups, and ask each one to take a different event that Levine describes. For each one, Levine provides a quick description of the event, his analysis of its significance, and selected quotations from his subjects. Each group should add its own spin to Levine's description, providing quotations of their own and taking a position for or against Levine's analysis. Discuss these revisions with the entire class.

Writer's Options

Levine describes the subjects of his 1979 study as having a "Titanic ethic"(7). Think of two or three metaphors that might characterize your generation, and use them to write a description of your generation. What aspects are most prominent in each metaphor? What aspects of your generation become more obscured when looking through a particular metaphor? After you have written full descriptions, pick the image that you think works best.

Multimedia Resources

Each of the events Levine's subjects identify has been the subject of a cover story for both *Time* and *Newsweek*. Bring in the cover illustrations (available at almost all libraries) and ask students to analyze the images in those photographs. Do they vary from magazine to magazine? What are the differences? What might those differences indicate about each magazine's position on each of these issues? What about the illustrations provided within the articles? Do they have analogous differences?

Two Perspective: Suggested Answers for Responding to Reading Questions

1. As a child, Bader had heroes who were larger than life, so to speak. They shaped history, they had guts, drive, vision, and dedicated followers. As an adult, she looks for role models, closer to those heroes that Levine's students describe. They have heroes closer to home, on a smaller scale, such as parents, grandparents, teachers, and unlikely anonymous folk who have done good deeds.

2. Answers will vary. Many students will probably mention the O.J. Simpson chase and trial, and the Oklahoma City bombing, the 1998 shooting of two Capitol policemen, and the Monica Lewinsky case.

3. Answers will vary.

4. Answers will vary.

5. Bader reverses the traditional meaning of "larger than life," showing that today, in her view, "larger than life" no longer exists; famous heroes can no longer just walk down the street without an entourage of secret service personnel, limousines, and police barricades. We actually get closer to these people through television and tabloid images. Life itself, then, is as large as it gets, and we need to find our heroes within that space. Levine would agree with this assessment, with students' attention to the material conditions that comprise their lives.

6. Answers will vary. Many students want to believe that both are possible, and will probably say so.

Using Specific Readings

"ELEGY FOR THE HOBBY," RANDY COHEN

For Openers

Ask students to list what they do in their leisure time. Are these activities primarily engaged in alone or with others? Ask students to focus on one activity that is particularly important to them. Does this activity reflect their family culture or their social culture?

Teaching Strategies

1. Cohen must establish criteria for what a hobby is so he can argue that present-day society does not conform to his criteria as hobbyists. List Cohen's criteria. Apply his criteria to board games, watching TV, and movie-going. Decide if Cohen's criteria are consistent and complete.

2. Notice the balance between Cohen's use of a humorous and a serious voice. Point out that his sense of humor and his many examples help to make his essay credible. Have students underline or highlight verbs and examples he uses to create his voice. At the end of the essay, Cohen becomes serious and explains the value of the need for hobbies. Hobbies do not only balance a day's activities, they also keep one mentally balanced and serve as a "defense against neurosis."

Collaborative Activities

1. Ask groups of students to review the school newspaper and bulletin boards. What characterizes the extracurricular activities of particular departments and of the school?

2. Compare hobby experiences in the group. Ask students, working in groups, to discuss to what degree they value those hobbies or hobbies in general. Do students have time for hobbies now? Do they think they will have hobbies after they graduate? Why or why not?

Writer's Options

1. Establish a set of criteria or an extended definition of what a hobby is. Argue that different examples of activities are or are not hobbies according to your criteria. Bring Cohen's essay into your paper for support or to refute your views. What are the implications or values of your criteria?

2. Argue that computer games or Monday Night football are comparable to other hobbies, such as board games and stamp collecting.

Multimedia Resources

Bring in sports, craft, woodworking, backpacking, and gardening magazines. Look at articles and advertisements in the magazines. Discuss how these advertisements promote competition or promote a healing leisure. Contrast today's magazines to magazines with similar topics from 30 years ago.

Suggested Answers for Responding to Reading Questions

1. Cohen calls hobbies a counterbalance to work. He gives several examples throughout the essay to show how, through time, different people have escaped the habits of work with hobbies. Cohen and his friends do not have hobbies because they are either bringing work home or grooming themselves for the next social event by working out or taking in the latest movie or cultural event.

2. When hobbyists make model sailboats, they are absorbed by the process of sanding, gluing, and painting. They are a intent on the authenticity of their replication. Beyond the crisis of gluing one's fingers together, there is no personal confrontation that would lead to transformation. Reading, on the other hand, is a vicarious confrontation; it throws us into crisis with the characters. Readers may experience a different culture through reading, and thus become more culturally sensitive. They may understand a physical principle, learn a new process, and thus grow. Reading has the potential to transform.

3. This is open to discussion. Some students may point out that technology provides different methods of diversion. Answers will vary.

Additional Questions for Responding to Reading

1. What does Cohen include under each of two categories of past times: hobbies that relax, and activities that help us "keep up?"

2. Are the examples Cohen offers from his own lifestyle typical of today's culture? What other examples can you suggest?

"LOSING IT," LAURA FRASER

For Openers

Ask students to respond in writing and then to discuss the following quotation:
"We will discover the nature of our particular genius when we stop trying to conform to our own or to other people's models, learn to be ourselves, and allow our natural channel to open."-- Shakti Gawain

Teaching Strategies

1. List people who may be considered modern-day heroes. What body types are represented? How many are overweight? Ask students to write the name of one of their heroes on a sheet of paper. Then, ask students to write why or why not their attitude would change if that person became 75 pounds overweight. Keeping the writing anonymous, collect the sheets and redistribute them. Invite students to read several of the papers. Discuss results.

2. Fraser ties success to thinness throughout the essay. Point out the parts of her essay that provide sufficient evidence to support and validate her assertion, "we stand before our mirrors in scrutiny, pinching our flesh, allowing our talents, achievements, and beauty to diminish before our size."

Collaborative Activity

In groups, discuss other ways that children and teens are expected to conform in order to gain acceptance not only from peers but from society as a whole. As a class, share findings and discuss possible advantages and disadvantages to both an individual and to a society when societal pressures to conform are imposed.

Writer's Options

1. Write an essay from the point of view of a visitor from another culture. How might that visitor describe the food rituals of your campus? What advice might they give? What conclusions can be drawn from their observations?

2. Write an essay in which you demonstrate how eating could be an enjoyable social event.

Multimedia Resources

1. Bring in a *Yellow Pages* directory. Ask a group of students to find diet and weight-related listings in your area. How big an industry is created around this perceived need to be thin?

2. Bring in slides or art history books showing Greek, Roman, modern European , Aztec, and African art work. How are the women portrayed? Are they thin?

Suggested Answers for Responding to Reading Questions

1. To be considered dating material, a female must have a thin body, so dieting becomes part of the rite of passage. This is reinforced through advertisements in popular teen magazines.

2. We all know what elevator music is. It is there, and we are barely aware of it until someone turns it off. Magazine racks, movie stars, models, manequins, create the distorted images of female perfection. Because these images are everywhere, they are easy to take for granted. However, the thinness message is pervasive. Consumerism may be another soundtrack being played. Great clothes, designer labels, and new cars are some examples of what one must have to be "in."

3. The every day attention spent on jobs and careers is a lot like her dieting experience. While both job and career occupy one's day, a career is a result of a choice of a mental habit; a career allows self-expression. A job allows that mental habit or self-expression to be acted out. A person can have a career in human development and hold many different jobs in that career.

Additional Questions for Responding to Reading

1. What is your response to the opening scene in this essay? Why is it important for the psychiatrist to have "domino-sized sideburns"? If you were in the office as his associate, what would you want to say to him?

2. How do you characterize the voice of the author in this essay? How does she try to relate to her reader?

"GETTING AWAY WITH MURDER," BARBARA GRIZUTTI HARRISON

For Openers

Discuss with your class the final sentence in paragraph 13: "if everyone is to blame, no one is to blame." Do they agree with this, or do they disagree? Why?

Teaching Strategy

One of Harrison's most effective stylistic techniques is the way she varies paragraph and sentence length. With your class, identify those places where Harrison inserts a very short sentence following a longer passage (for example, after paragraph 3, she simply writes, "So what.") How do your students react to these short sentences? Does this strategy catch their attention, or does it dilute what she is saying?

Collaborative Activity

Divide the class into groups, assigning one murder description to each group. Ask each group to record the type of information Harrison provides about each one. How was it done? What were the victims eating (or doing)? What did the perpetrator say? As a class, then, compare the types of information Harrison provides for each killing, and discuss why she chose the examples that she did.

Writer's Options

Write a letter to Harrison, explaining to her your reaction to her essay. Do you agree with her in some places but not others? Do you think she left anything out? What would you say to her if you had the chance? Do you think the "victim" claim is ever justified?

Multimedia Resources

The daily talk shows (*Oprah, Montel* and so on) devote specific episodes to the culture of victimhood every now and then. Tape one of these shows, and ask students to keep track of how many people—both on-stage and in the audience—seem to believe that those who have been abused cannot be held accountable for their actions. Keep track, too, of how many people seem to feel the way Harrison does, that we are responsible for ourselves regardless of bad upbringing or bad genes.

Suggested Answers for Responding to Reading Questions

1. Harrison uses similar casual language throughout the rest of the essay, as well: "the same sweetie pie" (7) and "[g]enes are big now" (15), to give just two examples. Some students will find this refreshing; others will find it annoying.

2. Answers will vary.

3. Harrison is saying that our society is skipping a step between conviction and forgiveness. Instead, we are tending to jump in to make the criminal feel better about him- or herself, without exacting any punishment or eliciting any genuine sorrow for the wrongdoing. People who know how to play the system will take advantage of this forgiving impulse, and they will continue to act without self-restraint or self-recrimination.

Additional Questions for Responding to Reading

1. Harrison relies on two authoritative sources, Cornel West and Martin E. P. Seligman. Do you find her use of these two men convincing? Explain your answer.

2. In the final paragraph, Harrison mentions the O.J. Simpson story. Knowing what you do now about its resolution, do you think Harrison was right to use it as an additional example? Do you think she assumed too much?

"LIFE IN PRISON," WEBB HUBBELL

For Openers

Ask students to explore through freewriting whether the American prison system should be dedicated to punishment, rehabilitation, or some other goals.

Teaching Strategies

1. Look over the structure of the essay. Hubbell puts himself in the situation of the prisoner. He then raises the different problems with the prison system, supporting each problem with some example. He moves on to assert that the prison system is not the fault of any one group, but a systematic problem of misunderstandings and erroneous solutions. He then goes on to propose solutions.

2. Hubbell contrasts what he knew as an outsider to what he learned as an insider. Go through the essay to show the organization of the essay. Note how transitions keep the essay balanced and coherent. A place to begin this analysis is with the repeating phrase, "I personally experienced." Ask students to highlight this phrase so they can see how Hubbell uses a repeating element in his essay.

3. Ask students to comment on the extended analogy with Sherman's march. Is it relevant? How does it help the essay achieve unity?

Collaborative Activity

Hubbell uses a variety of evidence. Ask groups to identify the different kinds of evidence (statistics, eyewitness accounts, interviews, anecdote, etc.). Then ask them to determine what kind of evidence is most persuasive to them and why. Finally, ask groups to describe how the evidence is woven into the essay, rather than just "dumped" in without interpretation.

Writer's Options

Hubbell says, "I never dreamed I would be locked behind bars," and "What I did not anticipate was being enlightened in a profound way."

Consider a problematic situation that you did not fully appreciate until you were an "insider." What solutions to the problem do you now offer? Consider using Hubbell's organizational strategy to help you organize your essay.

Multimedia Resources

The Shawshank Redemption, with Tim Robbins and Morgan Freeman, and *Bird Man of Alcatraz* with Burt Lancaster show how prisoners can use prison time for self-discovery. The essay "A Homemade Education" by Malcolm X (p.187) describes how, as a prisoner, he learned to read and write.

Suggested Answers for Responding to Reading Questions

1. Hubbell learned about the conditions of prisons, the effect of various policies, and the lack of education and retraining in prisons. He also learned about discrimination toward African-American males, and he also learned about new roles he might take as a former prisoner to improve prison conditions. Finally, he learned about his own sense of morality and had a chance to examine his own values. These lessons are permanent. Hubbell, early in the essay, admits his guilt and thus shows his remorse for the decisions he made. Answers to the last question will vary.

2. Both prison problems and his own self-assessment as a human and a citizen are exposed in this essay, there is a shared perspective on the personal and the general experience. Ultimately, he seems most interested in showing the prison's needs. All his personal experiences are related to show more clearly the larger issues.

3. By gathering first-hand reports from interviews with inmates and by studying statistics, Hubbell sees that the prison system is flawed. Other answers will vary.

Additional Questions for Responding to Reading

1. As you read this essay, what did you find disturbing or surprising in what Hubbell reported or in how you responded?

2. What does Hubbell claim is the purpose of the prison system?

"JUST WALK ON BY," BRENT STAPLES

For Openers

Ask students to recall a time when they felt fearful about a stranger. What was it about the person or the setting that provoked a sense of fear? What did they do? What do they now believe they should have done?

Teaching Strategy

Point out how Staples manipulates appearance and reality. He opens with a neat twist on the word *victim*. In the first paragraph, he speaks about himself as perceived by the woman — as a threat. In the second paragraph, Staples presents his own view of himself, "a softy who is scarcely able to take a knife to a raw chicken, let alone hold it to a person's throat." Use these observations to move the class into discussion of other words, such as *accomplice* and *suspect*, that Staples uses to contrast others' characterizations of him with his own.

Collaborative Activity

Have students work in groups to determine what traits cause different types of strangers to appear threatening. Encourage students to include signals that suggest deviance from social norms, such as modes of dress, facial expressions, behavior, and grooming rather than racial or ethnic identity. Then, in class discussion, explore the validity of these perceptions and point out the differences in what students find threatening. You might note, for example, that some people might be leery of young men with ponytails or earrings, features that many of your students will probably not find threatening.

Writer's Options

Write a response to Staples, identifying yourself either as one of the types he writes about or as a type of person he overlooked. If your point of view is different from his, how can you express it while remaining sensitive to Staples's position? If you feel the same way he does, how can you tell him about your own experience?

Multimedia Resources

In the spoof movie *Hollywood Shuffle* one of the recurring vignettes is of an acting school for African-American men. The joke is that they are all Shakespearean-trained actors who can only get roles as homeless men, pimps, or other assorted criminals. Show this brief scene to your students, and analyze the source of the humor. What makes it funny? What are the dominant values being expressed in this scene, and how are the actors poking fun at it? What might Staples say in response?

Suggested Answers for Responding to Reading Questions

1. Answers will vary.

2. One is left with the sense that breaking his habit of walking at night—being absent—is the only way Staples could be rendered harmless to the fearful people on the street. In this case, though, he would be doing all of the accommodating, which would be unfair.

3. Staples explains that "being perceived as dangerous is a hazard in itself" and points out the potential for death "where fear and weapons meet." Recent incidents in which African-American males have been detained by police because they "looked suspicious"—for example, because they were driving expensive late-model automobiles or jogging in white neighborhoods—show that Staples has good reason to be fearful.

Additional Questions for Responding to Reading

1. What does Staples mean by the "language of fear" (3)? What examples does he provide? Can you think of additional examples from your own experience?

2. How do you react to Staples's reference to Podhoretz's essay?

3. In what ways does Staples identify with the women he sees walking at night? Do you think this identification is justified?

4. Why does Staples repeatedly assert his harmlessness? Is this an effective strategy?

"ON DUMPSTER DIVING," LARS EIGHNER

For Openers

Eighner begins this essay with a lengthy exposition on the connotations of certain words: *scavenging, foraging, scrounging,* and so on. As a class, discuss the connotations and implications of the word *homelessness.* Eighner makes distinctions of various types of homelessness and he clearly distinguishes some types from others. Explore the many different types of homelessness in class discussion.

Teaching Strategies

Discuss the following with your class:

1. What is the essay's organizational principle? Eighner says he will start with specifics and move to more abstract ideas. Does he do this? What other principles govern the arrangement of his ideas?

2. What is Eighner's tone? How can you characterize it? If students wanted to replicate this tone, what stylistic strategies would they need to use?

3. What is the effect of all the details that Eighner uses throughout this essay? Do they paint a more vivid picture of his world? Or do the details help to obscure parts of his story that he isn't telling? Why does he use so many details?

Collaborative Activity

Eighner makes a joke about "scavenger ethics" although clearly there is an unwritten code of conduct among scavengers. In groups, ask students to search through this essay for evidence of a code of "scavenger ethics." Then, have them work as a class to pool all the examples they find and generate a list of rules.

Writer's Options

Think back to the last time you cleaned your room either at school or at home. What did you throw out? Make as exhaustive a list as you can. Then, write about the life that some of those objects might have taken on after you discarded them. Who might have taken them from a dumpster or landfill? What happened to them next?

To provide a little more background information on Austin, Texas, or large college towns in general, show students clips from the off-beat film *Slackers*. Set in Austin (where Eighner went to school), this movie is a series of random vignettes about those people who live within a college community without really being a part of it. If you teach at a major university, students might not be aware of this counterculture subsisting along with them. Eighner's homelessness and his articulateness fit in will with the characters in Slackers.

Suggested Answers for Responding to Reading Questions

1. Eighner seems to have multiple purposes here, which he achieves by saying he will instruct the reader in the art of Dumpster diving. Eighner's essay is a call for respect, as he shows that he is articulate and witty, and he makes distinctions between his class of scavengers and the haphazard winos who frequent the same areas. It is also a wake-up call designed to let readers know that a good education does not always preclude living on the street and living out of Dumpsters.

2. Answers will vary. Many will be surprised by Eighner's tone: amused and unapologetic. Others will find the science of diving, and the detail with which Eighner describes it, unexpected.

3. Answers will vary.

Additional Questions for Responding to Reading

1. Eighner writes, "I do not want to paint too romantic a picture. Dumpster diving has serious drawbacks as a way of life." Does he paint a romantic picture?

2. Eighner seems to have a clear classification system for people. What is it? Where would you fit in his scheme?

Focus: Where is Technology Taking Us?

The following three essays raise questions of how technology affects contemporary culture in the areas of personal relationships, communication, and sources that inform our views and beliefs. Ask students to list changes in technology that they have experienced in different phases of their lives.

"THE GLOBAL VILLAGE FINALLY ARRIVES," PICO IYER

For Openers

In paragraph 13 Iyer quotes a UNESCO official as saying "America's main role in the new world order is not as a military superpower, but as a multicultural superpower." Discuss what this means, and debate the accuracy of this statement.

Teaching Strategy

Analyze the structure of this essay with your class. How does Iyer organize his argument? What is his argument?

Collaborative Activity

Divide the essay into sections, each section describing a different type of global multiculturalism: goods, language, bloodlines, and so on. Ask your students to add to the list that Iyer includes, bringing in examples from their own experience. Are there major similarities and differences between Iyer's lists and their own? Why, or why not? Put the lists on the blackboard, and continue building on them as a class. Students can use these examples in their next paper, if applicable.

Writer's Options

Using as much detail as possible, write a description of your hometown, paying particular attention to multicultural influences. If you live in an urban area, choose a street to walk down mentally; if you live in a very rural area, choose the nearest town. Scrutinize the names on each storefront; recall the items that are for sale; picture the billboards and the ways in which they advertise. Can you find instances of multiculturalism that you never noticed before?

Multimedia Resources

Bring some maps into your classroom, and locate the places Iyer mentions on these maps. Are there parts of the world that he leaves out? Why might that be? Are there places that recur? Why is that?

171

Once you have annotated the entire world map with Iyer's examples, discuss how multicultural and inclusive his world view actually is.

Suggested Answers for Responding to Reading Questions

1. Answers will vary.

2. Answers will vary.

3. Many individuals feel a reduced bond with their families, and with a specific cultural heritage that their family members identify with. Others worry that the move to multiculturalism in fact eradicates difference, rather than celebrates it.

Additional Questions for Responding to Reading

1. Iyer begins this essay by calling himself a "relatively typical soul in today's diversified world." Do you agree with him? Is he typical? Why or why not?

2. In paragraph 2 Iyer writes, "I disembark in a world as hyphenated as the one I left." What does he mean by this? What would be another way of saying this? What other connotations does "hyphenated" have?

"PREGNANT WITH POSSIBILITY," GREGORY J.E. RAWLINS

For Openers

Ask students what their experiences have been with the Net, particularly in reference to chat rooms and e-mail. How has their correspondence via the Net been positive and negative? How can this correspondence become dangerous?

Teaching Strategy

Rawlins's essay assumes that the reader is familiar with computers and with surfing the Net. Ask students to fill in examples for his general references. For example, "As it quickly commercializes, and changes drastically..." What are examples of this commercialization?

Collaborative Activity

Divide the class into three groups to exchange ideas about a particular topic. One group will have time to talk during the week. Another group will carry out an e-mail exchange. The last group will send snail mail (regular postal system). Examine the results. Do shy people talk more on-line than in discussion groups? What is the volume and quality of the exchange?

Writer's Options

1. Interview several people about their use of the Internet and e-mail. Ask how frequently they use it, what purposes this kind of communication serves, and how it affects their personal relationships. Interpret and evaluate the results.

2. Engage in an active e-mail exchange (3-4 per day) with someone in your class. After one week, reflect on what you have noticed and write a response to Rawlins's essay.

Multimedia Resources

1. "Listen in" on a professional listserv. Notice the language used and the length of the exchanges. Contrast these listserv exchanges with non-professional chat rooms and responses.

2. Go to the library in person to research a topic. Then, research the same topic on-line. Compare the information you were able to find and the social contacts you made.

Suggested Answers for Responding to Reading Questions

1. In the beginning of the essay "Only connect" refers to intimate human relationships. The final "Only connect" suggests accessing people or ideas via the Internet. Iyer's essay shows how cultures are absorbing each other's traits; we are connected by belonging to multicultural communities, but we can become very isolated.

2. Answers will vary. One could argue that information accessed on the Net could be obtained in other ways. Conversely, information and opportunities can be announced more quickly over an electronic server than through printed text.

3. Rawlins's attitude is positive and at times even enthusiastic. Gup and Rawlins would agree that computers can change the way we communicate. Rawlins and Iyer would agree that the values of communities are becoming more permeable and that there are more opportunities to interact with people beyond one's own geographic boundaries.

1. According to Rawlins, despite any technical advancement, economic classes will always govern the distribution of opportunities. Do you agree?

2. In what ways has the technology that Rawlins describes affected your own college life?

"THE END OF SERENDIPITY," TED GUP

For Openers

Ask students to freewrite, and then discuss their attitudes and experiences with "doing research." See what their negative and positive attitudes and experiences can be related to what Gup is driving at in his essay.

Teaching Strategies

1. Gup uses several metaphors to describe his thoughts, such as "information was a smorgasbord," "the mouse is a key, it is also a padlock," "cyber-hermits," "yet another moat," and "there are few Luthers." Spend some time explaining the metaphors. Show how these various references demonstrate Gup's eclectic familiarity with a variety of topics and expressions.

2. Gup suggests that research is a tactile experience so that our encounter with textual material progresses "first into [our] hands and then into [our] minds." Ask students to consider how they learn and acquire understanding in their classes. How much of their learning process is tactile? When they do work on computers do they miss "a cumbersome front page"?

Collaborative Activity

Ask students to bring to class several newspapers and magazines. Ask the groups to test Gup's claim that by having texts in our hands, we are more likely to engage with those who are "on the margins of our consciousness." What draws our eyes to articles?

Writer's Options

1. Has your experience with computer research been as efficient and direct as what Gup describes? Is it possible to wander and

be caught by an article extraneous to your original target in cyber-searches? Give specific examples of how serendipity may still be experienced with technology.

2. Describe and analyze your experience with using on-line texts and Web pages versus books and pages you can hold, scribble on, and thumb through. What moral and ethical evolutions should we embrace and resist as we move toward distance learning, on-line classes, and on-line discussions?

Multimedia Resources

Browse any topic on a search engine such as Metacrawler or Yahoo. How much browsing can be accomplished on-line? How much browsing were you able to do on other non-related topics as a result of your initial search? Go to your school's library Web Page. Select "search engines." Compare the university's offerings to the Net's offerings.

Suggested Answers for Responding to Reading Questions

1. Gup feels that without the opportunity to browse through a variety of topics and become distracted by articles that may tease our curiosity, the serendipity, the happenstance of discovering new ideas and information, will be lost.

2. Because we are members of a global community, we are morally bound to be aware of conditions beyond our own limited experiences and needs within our local community. Therefore, the moral consequences of not knowing the conditions and "news" of the marginalized will isolate us and will isolate them. We have a moral imperative to put ourselves in situations where our complacency and ignorance of global events is challenged.

3. Answers will vary.

Additional Questions for Responding to Reading

1. How does Gup's vocabulary affect your reading of the essay?

2. What part do reflective practices play in your own education?

CHAPTER 8

MEDICAL PRACTICE
AND RESPONSIBILITY

Setting Up the Unit: Using Specialized Language in Order
to Identify with a Particular Community

The readings in this chapter were written mostly by either members of the medical community or people involved in related professions (like Jane Goodall). In most cases, the writers identify themselves as experts by using technical terms from their professions. For example, both Abraham Verghese and Richard Selzer use terminology that is probably only understood by those trained in medicine. Likewise, Jane Goodall relies on phrases from the behavioral sciences. In other words, by using the specialized language of a certain community, each writer identifies him or herself as a member of that field.

As you can see, language choices that identify a writer with a certain group do not necessarily make the writing incomprehensible. In fact, it is still easy to follow Selzer's ideas without knowing exactly what those medical terms mean; those phrases are used more to show his knowledge and training than they are to convey a specific meaning to readers. In the case of Goodall, the words used to identify her within her primary area of interest are also accessible to a general reader. Specialized language choices not only indicate the community to which the writer belongs but also indicate the writer's point.

Students, too, can work with specialized language. Each individual belongs to a range of communities, each with its own conventions for appropriate discourse. By selectively using the language of a given community in their writing, students can either show readers that they are in the same community, or they can bridge the distance between a reader's community and their own. Many times, however, students choose not to use words that they think don't fit in an academic paper, writing instead a version of academic English that lacks personality. Introduce them to this idea, and encourage them to use language that conveys the range of various discourse communities to which they belong.

In some cases, of course, students have the opposite problem. Some need to learn not to use so much of their other vocabularies. If this happens in your class, discuss instead the concept of *jargon*. When specialized language obscures meaning and alienates the audience, it is no longer effective. At this point, call it *jargon*, specialized language that excludes vast numbers of readers from understanding a piece of writing. Jargon, when used appropriately, communicates to the audience that you are a part of their club; when used inappropriately, it communicates just the opposite.

Confronting the Issues

Option 1: Constructing Contexts

Very few aspects of our lives are not entwined in some way with the medical community, or with medical issues. Indeed, many contemporary "great moral debates" deal with issues of medicine and health care. In order to have students discover this fact on their own, bring a week's worth of newspapers into your class, and ask students to identify all of the articles that have something to do with health care or with medical issues. After generating a list as a class, each student should pick a topic from the list that seems most compelling to him or her, and write a paragraph to the class explaining why. Read or distribute these paragraphs to the rest of the class, grouping students with similar interests, if possible. What issues suggest the most questions for your class? Which questions are worth pursuing for a major project? Why? Discuss any trends you see in *their* interests, encouraging a full class discussion. Possible topics include:

- AIDS
- Animal testing
- Alternative medicine
- Euthanasia
- National health insurance
- The role of the Surgeon General

- Secondhand smoke
- Cancer research
- Overpopulation
- Alzheimer's disease
- Nursing homes
- HMOs
- Sudden Infant Death Syndrome
- Medical school training
- Organ donation/transplants
- Birth control
- Plastic surgery
- Physical therapy
- Paralysis
- Pharmaceutical sales
- Living wills
- Treatment of alcoholism/drug addiction
- Fetal tissue research

Option 2: Community Involvement

This chapter raises issues about the ways in which Americans depend upon the medical community, and several readings explore shortcomings associated with modern medical institutions. Your class can research alternatives to traditional medicine that are available in your community. What alternatives to traditional maternity wards—for example, birthing centers, midwifes, or resources for home births—exist? Likewise, what sorts of hospice organizations exist? How about alternative medicines? Herbalists? After conducting research in community libraries, social service agencies, counseling centers, and health clinics (just for starters), students can collaboratively compile a resource manual that is specific to your community, with useful information for potential customers. When it is complete, they can distribute this information through a local library.

Option 3: Cultural Critique

The chapter presents a good opportunity to examine the myths and images surrounding the role of medicine in contemporary American society. First, ask students to briefly stereotype a doctor, then come up with a list of the personal qualities and professional responsibilities that such a person is supposed to have. Next, students should write about an experience they have had recently that somehow involved a medical transaction (for example, a visit to the gynecologist, treatment for a broken arm, a visit with a friend who is going through chemotherapy). Ask the class to write their lists on the blackboard, noting any overlap. Next, ask them to compare the reality of their recent transactions with the stereotypical image of the doctor. In what ways do they match? In what do they not match? This exercise should provide a good springboard for discussing the differences between our expectations of medical professionals and the care consumers actually receive.

Option 4: Feature Film

Show the film *Medicine Man* to your students. Many of the issues that arise in the readings throughout this chapter (and in the next chapter, as well) run through this movie. It is about a young, female medical researcher from New York City (similar to Perri Klass) who joins a reclusive doctor in the wilds of the Amazonian rainforest. The doctor has been hiding from society after his call for additional research helped introduced the swine flu to the village he had lived in, killing the entire population. Now, the same doctor thinks he has found a cure for cancer. Ask students to watch the movie to see how the issues of medical responsibility are addressed.

Teaching "Two Perspectives"

Before reading either of these essays, ask students to consider their reactions to the student voices that open this chapter. Do any of the student perspectives on AIDS or death make them angry? Uncomfortable? Do they agree with any of them? Ask them to write a response to one of the students in which they tell why they either agree or disagree with something the student has written. Then, as they read the Verghese and Klass essays, ask them to write a similar response to one of those two authors on the same subject as their response to the student.

"MY OWN COUNTRY," ABRAHAM VERGHESE

For Openers

Ask students to explain the type of AIDS education they received in school. Some will have heard quite a bit; others may have heard of it only on the news, as a "big city problem." Compare the types of programs students identify with the places in which they grew up. Is there a pattern? Do differences seem to be based on location? On population? On students' ethnicity? Do your students still have any misconceptions about the disease, and how it is transmitted?

Teaching Strategies

Discuss the following with your students:

1. Verghese shifts his perspective throughout the essay, putting himself in others' shoes at different points. Identify these shifts to your students, and ask them to discuss how they react to them. Does the narrator's voice make them more sympathetic to the young man? To the doctor? To his colleagues?

2. Between paragraphs 5 and 6 Verghese abruptly shifts from describing scenes of natural beauty along the road to describing the high-tech world inside the hospital. Ask students what he gains with this contrast. Why does he begin the essay with the lengthy description of the car ride?

Collaborative Activity

Verghese uses a number of techniques to make the patient more than a faceless, nameless case. Divide students into groups, assigning a section of the essay to each group, and ask them to identify the ways in which Verghese humanizes the man with AIDS. Keep a running list on the board, and see which techniques are common throughout the essay, and which ones occur only at certain moments.

Writer's Options

Retell the story from the perspective of one of the people Verghese mentions in paragraphs 42 through 47. How do their stories differ from Verghese's version? What might they notice, and relate, in retrospect?

Multimedia Resources

Bring in a tape of an opening scene from the television show *ER*. (Almost every episode begins with a scene in the emergency room.) Ask students to pay particular attention to the type of care the patients receive as soon as they arrive in the emergency room. What kind of information do the doctors have at their disposal? What kinds of decisions do they need to make immediately? What precautions do they take? Reread "My Own Country" in light of the kind of care that emergency practitioners have to give.

"INDIA," PERRI KLASS

For Openers

In paragraph 11 Klass cites the reason for sending American medical students to developing countries: to learn about diseases they would never see in the United States because we have already eradicated them. Discuss with your class the moral and practical implications of this kind of training. Is there anything wrong with it? Should young physicians stay in the United States and work in areas that have severe shortages of doctors? Should they be trained in other ways instead?

Teaching Strategy

List all the cultural misunderstandings Klass experiences. What implied criticisms of Indian culture are inherent in Klass's misunderstandings? Likewise, what implied criticisms of herself are inherent in her assumptions?

Collaborative Activity

Divide the class into groups, and ask each group to read through the text to find the contrasts that Klass describes. Make two columns, one for life in India and one for life in Boston. Keep corresponding details together, matching positive traits on one side with negative traits on the other. Does Klass identify anything positive about life in India? Can your class derive any lessons for themselves from the way Klass simply breaks things down into good and bad?

Writer's Options

You don't need to be in another country to misread a situation because of a cultural bias. Write about a misunderstanding you had, or you witnessed, because of cultural barriers. What happened? What was the bias? How did it get in the way? What did you learn from the situation?

Multimedia Resources

For some additional background, and for visual detail, bring in some scenes from the film *City of Joy*, in which the lead character encounters a culture shock similar to the one that Klass describes.

Two Perspectives: Suggested Answers for Responding to Reading Questions

1. Verghese describes the patients time in the hospital in order to humanize and dehumanize him. By showing all the treatments that the man underwent, Verghese conveys the desperate and serious nature of his affliction. By recording all these treatments in medical language, Verghese shows how the man is just another case to be treated until his AIDS diagnosis sets him apart from other nameless cases. The dual tensions of humanizing and dehumanizing run throughout this essay.

2. The medical staff in Johnson City reacted with fear when they found out that the patient had AIDS; not very much was widely known in 1985 about how the disease spread—and did not spread. Mostly, the staff was concerned for itself, and a little embarrassed that they had been seemingly duped. It would be nice to think that they would respond differently today, since there is less hysteria and more understanding surrounding AIDS patients, but in some pockets of the country, they might still react the same way today.

3. Klass does not understand the culturally determined aspects of Indian appearance, speech, and clothing. She is unfamiliar with the different environment and diseases, and with the different expectations and attitudes. Finally, she sees terminally sick patients as unusual and unfortunate; the Indians see death as an unavoidable and natural occurrence. Some of these differences are the result of reduced access to health care; others are ingrained in the cultural and religious beliefs of the people of India.

4. Both Verghese and Klass focus more on the medical and support staff (including family members) than they do on the patients. And both are upset by what they perceive to be a lack of concern for the individual patients involved. Klass's disappointment, however, stems from economic and cultural circumstances outside her control, while Verghese's disgust is rooted in the social stigma attached to homosexuality and AIDS.

5. When death is seen as inevitable, people tend not to fight it. Klass, in particular, sees too many people just giving in without challenging death. Doctors often feel they need to enlist the patient in the effort to stave off death.

6. Both writers assume the voice of the American medical community, so it is easy to identify with the doctors. Verghese, however, tries to put the reader in the mindset of the patient as well, to get us to see the situation through his eyes as much as possible. Klass, on the other hand, seems to appeal only to Western readers who share her assumptions about the role of medicine in our lives.

Using Specific Readings

"IMELDA," RICHARD SELZER

For Openers

Discuss the issues surrounding plastic surgery with your students. Why is it such a big industry in the United States? When is it a necessity? When is it vanity? What kinds of people generally have access to plastic surgeons, and for what purpose? Who else should have this access? What about burn victims? Cancer victims in need of reconstructive surgery? Children like Imelda, with congenital deformities?

Teaching Strategies

Trace Selzer's relationship with Dr. Franciscus. How does he characterize it at different places in the essay? How does it change? Ask students how Selzer shows the ways in which their relationship has been altered.

Collaborative Activity

In paragraph 9 Selzer describes Franciscus as "a man of immense strength and ability, yet without affection for the patients...[i]t was less kindness that he showed them than a reassurance that he would never give up, that he would bend every effort." Divide the class into groups, assigning a different segment of the essay to each group. Ask each group to find evidence to support Selzer's assessment. Are there times when Franciscus seems more kind than Selzer realizes? Are there places where it seems Selzer's assessment might be wrong?

Writer's Options

Write a description of somebody you admired who has died. Selzer's treatment of Dr. Franciscus is honest, kind, yet critical in places. Strive for the same balance, giving your subject credit for what he or she brought to your life, while acknowledging any shortcomings that might have seemed contradictory.

Multimedia Resources

Ask students to compile images of doctors from various sources. They can bring in videotapes, cartoons, posters, magazine articles, advertisements, movie clips, textbook photographs, and so on. You can supplement their offerings with some historical renderings of doctors. Ask your students to articulate the dominant images, and discuss how Selzer and Franciscus correspond to those representations.

Suggested Answers for Responding to Reading Questions

1. The narrator provides two reasons for the post-mortem surgery, the first of which is that the doctor was compelled by the mother's grief. Later, however, Selzer seems to feel that fixing the girl's lip was also an act of compassion for himself, to show that he could see patients as more than their pathologies and numbers on charts.

2. Answers will vary, depending on the associations that students have had with doctors, and depending on the representations of doctors that they follow on television and in the movies.

3. In many ways, this is the story of Richard Selzer finding out for himself what he wants — and doesn't want — to be when he

grows up. At first, he is in awe of Dr. Franciscus, and he looks to his elder as a role model. As the story progresses, however, Selzer distances himself more and more from the strictly scientific approach that Dr. Franciscus appears to take.

Additional Questions for Responding to Reading

1. Compare Selzer's treatment of Honduras with Klass's treatment of her setting in "India." What are the similarities? How does Selzer's handling of his patients' poverty level differ from Klass's?

2. What is Dr. Franciscus's secret? Why does Selzer feel so compelled to keep it for him?

"THE BLACK DEATH," BARBARA TUCHMAN

For Openers

This essay traces the devastating effect of bubonic plague on fourteenth-century Europe. Ask students how a documentary film about the plague could best inform and involve viewers. Would it have a voice-over narrative of Tuchman's words? Readings from journals? Illustrations? What meaning might such a documentary film have of contemporary views?

Teaching Strategy

Discuss whether our advanced science and technology would make us immune from the hysteria that accompanied the bubonic plague, or whether technological advances such as television and radio would make us even more panicky.

Collaborative Activity

Ask different groups of students to rewrite a portion of he essay in different ways –as a contemporary diary account or official report, a twentieth-century medical analysis of the disaster, as a descriptive poem. Let students present their creations and discuss how their plague accounts differ from Tuchman's.

Writer's Options

Tuchman writes, "Human behavior is timeless." If such a plague struck modern society, do you believe people today would abandon their children and spouses as men and women did in the fourteenth century? Write your response to this question. You might want to discuss the Ebola Virus, which struck parts of Africa.

Multimedia Resources

There have been a number of popular films recently addressing the issue of sudden, untreatable plagues. Two example are Stephen King's made-for-TV movie *The Stand* and the feature film *Outbreak*. Using one of these as illustrations, ask students to discuss modern treatments of plagues, as well as reactions to those who are infected. Consider also why these films are so popular.

Suggested Answers for Responding to Reading Questions

1. Since Tuchman considers herself "a writer, whose subject is history," then it follows that to create characters, tension, and settings, she will use descriptions that will involve the reader both logically and emotionally. History is more than the reporting of events and dates; it is the story of those events. She uses details to help the reader understand the reality of the plague.

2. Answers will vary. The most compelling supporting details are often those that help us understand the immediate reality of people living through the plague and those that help us identify with their plight. Quotations from modern sources would have undercut the immediacy and power of Tuchman's description.

3. The temptation to color the presentation of historical fact with sentimental language, to rhetorically arrange the information for dramatic effect, is great. Another writer might subjectively recreate the suffering of patients, the mass hysteria, and the general sense that the end of the world is near. And this type of treatment could be powerful for some readers. Tuchman, however, chooses to present the stark brutal reality of the plague. Her austere, detached treatment effectively keeps modern readers in the world she describes.

1. Would you call AIDS a contemporary form of the Black Death? Why or why not? In what way are those two diseases similar? In what way are they different?

2. Does Tuchman express any personal opinion in this essay, either directly or through her selection and arrangement of facts?

3. Why does Tuchman include a discussion of the great plague of Athens in 430 B.C.?

4. How did people account for the disaster in the fourteenth century? How might the twentieth-century explanations differ if a similar disaster struck?

"A PLEA FOR THE CHIMPS," JANE GOODALL

For Openers

For the past decade or so, more and more health and beauty products (shampoo, make-up, lotions, etc.) have been marketed as being free from animal testing. Discuss this trend with your students. Is it just a marketing ploy? Have students ever bought one product instead of another because it didn't rely on animal testing? How much of an issue is animal testing in their own purchasing habits?

Teaching Strategy

Goodall structures her argument very carefully, always considering her relationship to the chimps as well as her relationship to potential readers. Analyze the progression of her essay with your students. How does she make her position seem both rational and middle-of-the-road? How does she establish her credibility? Against whom does she set herself?

Collaborative Activity

Throughout the essay, Goodall uses a number of techniques to link chimps with humans in terms of behavior, physiology, emotional life, and potential intelligence. Divide your class into groups, assigning a section of the text to each group. Each group should generate a list of the ways—subtle and overt—that Goodall compares chimps with humans. Keep a running list on the blackboard, and see how many different strategies she uses.

Writer's Options

Perhaps surprisingly, Goodall shows quite a bit of empathy for the caretakers, calling them "victims of a system that was set up long before the cognitive abilities and emotional needs of chimpanzees were understood" (28). Posing as Goodall, write a letter to one of the caretakers, suggesting steps that he or she could take to alleviate the problems that confront the chimps.

Multimedia Resources

Bring to class, or ask your students to collect and bring to class, posters, buttons, pamphlets, and newsletters distributed by any of the various animal rights groups. How much variation is there in the types of graphics they use? Do they seem to be addressing different audiences? What are the major persuasive tactics each group uses? Compare these strategies with those that Goodall uses; which are more effective in your students' opinions? Why?

Suggested Answers for Responding to Reading Questions

1. Goodall says that she won't address the ethics of using animals for research as a technique to build rapport with her audience; it makes readers listen more closely when she focuses on what seems to be a less controversial issue. But by setting up her essay with the larger question of the ethics of any animal research, she leaves the issue in readers' minds. Furthermore, by speaking for the chimps, as she does in the final sentence, she subtly reintroduces the issue.

2. Answers will vary, but most students will mention the financial investment necessary to equip their labs in the ways that Goodall suggests.

3. Answers will vary, although many will say that Goodall would agree not to use animals at all.

Additional Questions for Responding to Reading

1. In paragraph 35 Goodall writes, "If we do not do something to help these creatures, we make a mockery of the whole concept of justice." What does she mean by this statement? Has she

talked about justice previously? What assumptions is she making about her readers at this point?

2. Would this argument be as effective if it were coming from somebody who does not have Goodall's background? How much credibility does her field research give her? Does it undermine her argument in any way?

"MY WORLD NOW," ANNA MAE HALGRIM SEAVER

For Openers

What can be done to make life in nursing homes more bearable? If some students have volunteered at a home, or have relatives who live in one, they can offer examples from their own experience. See if your class can generate a list of ways in which Seaver's life might have been improved in her nursing home.

Teaching Strategy

Ask students to consider who Seaver's audience is. Who might she have had in mind as she wrote? What was her purpose in writing? Might she have hoped that someone in particular would find her notes and read them?

Collaborative Activity

One of the qualities that makes this piece so compelling is the touch of humor that Seaver injects when she writes about certain issues. Ask students, working in groups, to find places where Seaver uses humor to make a serious point. Then, students can rewrite the passage using a more angry tone. Which version is more effective? Why?

Writer's Options

Compose a journal entry as if you, too, were in the nursing home with Seaver. What would you do all day? What would you write? How would you reach out to people, if at all? What would you like to see changed?

Multimedia Resources

Bring in a video of the Pepsi commercial that shows the "old folks home" that is energized when a delivery truck brings a load of Pepsi (and, conversely, the fraternity house that begins to play card games in rocking chairs when they receive a shipment of Coke). What is the image of the elderly in this ad? Consider, too the way the elderly are portrayed in other commercials. What products are pitched to them?

Suggested Answers for Responding to Reading Questions

1. The inconsistent structure evokes the feelings that Seaver traveled through in the course of a day. She has moments of humor, of anger, of dignity, of compassion, of irritation, of impatience, of docility. She probably wrote this in snatches and like many ill older people, she probably has trouble concentrating and her mind probably wanders. A more coherent structure would seem artificially imposed.

2. Seaver creates the picture of life in a nursing home as lonely, humiliating, and uncomfortable. She describes her annoyance with her lack of control of her body and of her daily routine. The residents at the nursing home are characterized as formerly successful people who were leaders in their own worlds.

3. Most likely, Seaver's son felt guilty about leaving his mother in the home, just as he probably regretted that he could not visit her more often. He could have published this essay to make himself feel better, but it seems that he put her words together in order to let other adult children know what their elderly parents' lives are like in the homes.

Additional Questions for Responding to Reading

1. Seaver makes repeated links between raising her own children and her situation in the nursing home. Why? What effect does this have?

2. In "Limited Seating on Broadway" (Chapter 6), John Hockenberry discusses some of the same physical discomforts that Seaver does in her notes. What do the two writers have in common? What might they say to one another, given the chance? In what ways are their circumstances very different?

"ON THE FEAR OF DEATH," ELISABETH KÜBLER-ROSS

For Openers

Discuss the final paragraph with your students, asking them to respond to the question that Kübler-Ross raises: "Are we becoming less human or more human?" Students can draw on their own experiences with deaths of loved ones as well as on the information that Kübler-Ross provides in the essay.

Teaching Strategy

If discussion is slow, you might want to ask students to freewrite about a time when a death touched their lives. They can then share their writing with a group, and the group can decide which group member's writing to read aloud to the class. After the personal writing has been read, students may be more willing to talk.

Collaborative Activity

Ask each group of students to write and prepare for discussion a detailed hypothetical case of a critically ill person (it may be based on a true story, if they prefer—perhaps Christopher Reeve). Discuss as a class how the patient's situation might have been handled. Topics that should be considered might include euthanasia, "death with dignity," hospices, life-support systems, organ donation, brain death, and so on.

Writer's Options

Write a living will, giving instructions to your family about the circumstances under which you would (or would not) like your life prolonged by artificial means. Consider as many of the topics generated for the *Collaborative Activity* assignment as you wish.

Multimedia Resources

Read aloud the scene from Aldous Huxley's *Brave New World* in which the children are indoctrinated to be unafraid of death. How does this attitude toward death compare with that of other cultures that students know of? If they are interested in learning more, they can turn this into an informal research project.

Suggested Answers for Responding to Reading Questions

1. Answers will vary.

2. Answers will vary. Personal experience will strongly influence opinions and counterarguments.

3. Anecdotes underscore the human rather than the mechanical approach to death; they may also reflect the reader's personal experiences and thus may be more persuasive than scientific support. If Kübler-Ross were to address an audience of physicians, however, she would probably decide to add scientific data to her essay.

Additional Questions for Responding to Reading

1. Kübler-Ross says that many customs and rituals concerning death may be traced to our own guilt and fear of punishment. Do you agree? If not, to what do you think they might be attributed?

2. Do you believe that meeting a critically or terminally ill patient's psychological needs should take precedence over meeting his or her physical needs? Explain your answer.

3. Compare the way Kübler-Ross characterizes attitudes toward death with the way Perri Klass discusses them in "India" (earlier in this chapter). How do the two writers differ in their views toward death? Why do you think they have such different approaches?

"WHAT NURSES STAND FOR" SUZANNE GORDON

For Openers

Ask students to write as a list or as freewriting, all the words or phrases that they associate with the nursing career. After they have generated the list, ask students to write a brief statement of what their list suggests about this field of medicine. Then discuss how their preconceptions of this field fit in with Gordon's essay.

Teaching Strategies

1. Just as hospitals do, many students have to budget their money and manage their college expenses. Discuss where most of their money is spent. If they have to make some

budget cuts, what do they really cut first? What would they cut last?

2. Does everyone agree that nurses cannot be easily replaced and that their role as caregivers to both body and mind is essential? Is Gordon overstating the irreplaceable position of nurses?

3. Teaching has fallen under the hatchet of downsizing, resulting in overcrowded classrooms and tired teachers. To what degree should schools and hospitals be run like corporations? To what degree should they not be run like corporations?

Collaborative Activity

Play a game of "Lifeboat." Divide the class into groups with five members each. Each group is in a lifeboat. Have prewritten on index cards these titles: Hospital Administrator, Surgeon, Medical Technologist, Nurse, Nurse's Aid, Unlicensed Assistive Personnel, Processor of Insurance Policies. Set the scene that each lifeboat has been stranded and there is very little food and water available. Each day, it has been agreed, one person will have to swim away for help, with very little chance of survival, so that there is more food and water for the rest. Eventually, the person whose life is most important will be left in the boat. Ask students to argue who should go first, second, third, etc., giving specific reasons. The person who is going to go overboard should provide a counterargument.

Writer's Options

1. Write an informational essay that shows the quality of education for nurses. Check online catalogues of universities to compare programs. Make a case that the expertise of the nurses is a necessary component in caregiving; show how their education is more than "training" in bedpan care and sponge baths.

2. Interview nursing students at your school and professional nurses in your community. What motivates them as professionals? What hopes do they have for their careers? Write a persuasive essay on some angle you have found developing in your conversations. Be sure to refer to Gordon's article in your essay.

Multimedia Resources

1. Since *ER* was referred to in the essay, tape a few episodes of the show, and ask students to describe the various roles played by nurses. Are they depicted as professionals or just warm fuzzy women?

2. Check the "National League for Nursing" home page: http://www.nln.org for information and viewpoints on issues and working conditions.

Suggested Answers for Responding to Reading Questions

1. The narratives at the beginning of the article help put the reality of the issue into the foreground, showing nurses in action with real people. Many readers will be able to identify personally with the action. These stories demonstrate the intricacy and intimacy of the profession.

2. Gordon shows that UAPs are non-medical personnel who care only for the physical needs of the patient. Because these untrained workers cannot read signs of general well-being or illness, the overall care of the whole patient is neglected or incomplete.

3. Gordon holds that the intimate nature of the nurse-patient relationship is not part of open conversations; therefore, the work done by the nurses is not openly recognized or analyzed. What the media popularly disseminates is the notion that nurses clean up after patients, assist the doctors by obediently or mindlessly following directions, and just being around. This view demeans and diminishes the professional nature of this highly trained profession.

Additional Questions for Responding to Reading

1. As you read this essay, what images of the nurse do the stories and the descriptions create for you? From your experience are these realistic depictions of nurses?

2. Based on this essay, how would you write a job description of hospital nurses?

Focus: Whose Life Is It Anyway?

In these next three essays students will confront different views on the right to die debate. Each author tries to identify what is at the heart of the debate. Ask students to note that each author refers to some personal connection to the issue. Ask students to be sensitive to specialized vocabulary of the two doctors and one lawyer. Also, ask students to watch for subtle connotations made in each essay.

"THE ETHICS OF EUTHANASIA," LAWRENCE J. SCHNEIDERMAN

For Openers

In paragraph 17 Schneiderman makes a connection between the debate surrounding abortion rights and the arguments about euthanasia. Discuss the similarities and differences with your students. How are the two debates alike? What issues do they share? Where do they depart from one another? At what point do arguments for or against one not serve the other?

Teaching Strategy

Schneiderman introduces the idea of an "ethics consultation" in paragraph 9. Discuss this idea with your students. Such consultations are becoming increasingly common in medical practice, and most medical training now includes medical ethics. With your students, generate a list of other institutions — for example, the legal profession — that could benefit from a formalized "ethics consultation," which would consider issues on a case-by-case basis.

Collaborative Activity

After dividing the class into groups, ask each group to take one of the questions that Schneiderman poses in the final paragraph. Students in each group should discuss the issue among themselves, and then lead a class discussion on that specific question. What issues does it raise? Can your class come to a consensus on how any of these questions should be answered, or are they unresolvable questions?

Writer's Options

Schneiderman mentions several different kinds of death: "easy, pleasant death," "ugly, debasing death," and "good death." Write your own definition of a "good death," contrasting it with a bad one.

Multimedia Resources

Schneiderman mentions the hit play *Whose Life Is It Anyway*, which was made into a film. If your students are interested, rent a copy of the film and show it to the class. Do they agree with Schneiderman's assessment of it? Does the film make the eventual death seem uncomplicated and clean? Or does it deal more directly with the physical unpleasantness and uncontrollability of dying?

Suggested Answers for Responding to Reading Questions

1. Answers will vary.

2. Schneiderman is saying that public debates about issues like euthanasia often center on universal decisions, arguing about whether an act is right or wrong—always. His case-by-case approach takes particular patients and their individual situations into account, allowing for more informed and more appropriate decisions to be made. People tend to debate ideals more than specifics, which Schneiderman shows can be messy and unpleasant.

3. Schneiderman does not resolve the question of euthanasia with the same assurance and confidence that Kevorkian displays. Schneiderman shows through questions on different cases and situations, that as a community we must decide on our mores and procedures. He would not agree that one person can act ethically on an issue that must be resolved *both* personally and socially.

Additional Questions for Responding to Reading

1. What is Schneiderman's tone throughout this essay? Does it shift at any point? When? Why? Does his tone in the first paragraph make you receptive to his argument?

2. Schneiderman spends several paragraphs providing examples and cautionary tales from different historical perspectives. What does he accomplish with this strategy?

3. Do you think that Schneiderman's case-by-case approach lends itself to legislation? How do you think he would vote in a referendum on euthanasia? Support your answer with evidence from the essay.

"A CASE OF ASSISTED SUICIDE," JACK KEVORKIAN

For Openers

You might need to begin by bringing Kevorkian's legal status up to date. In paragraph 3 Kevorkian writes, "after all, it is one's mental status that determines the essence of one's existence." Debate this point in class. What might others say determines the essence of one's existence?

Teaching Strategy

Kevorkian is careful throughout the essay to build his credibility. Trace the ways in which he does this. Examples include: screening his applicants; providing criteria for the "ideal" candidate; aligning himself with professional communities; setting himself above the rest of the medical community; showing his efforts to be honest about his purposes; including the patient's wishes throughout the essay.

Collaborative Activity

Divide Kevorkian's account into discrete stages, from the initial contact by Janet and her husband, to obtaining the right information on her state, to securing a place for the procedure, to the final act of "medicine." Ask each group to consider whether or not they think that Kevorkian did a credible job at each of these stages. Each group should try to come up with a position, however tentative. Chart the stages on the blackboard, and see where (if at any point) the class feels Kevorkian's actions were irresponsible.

Writer's Options

Kevorkian justifies his choice of patients in several ways. Do you agree that Janet was a good choice? If so, write a case study of her explaining why. If not, compose a profile of a patient who would be a more appropriate choice.

Multimedia Resources

A number of magazine articles and newspaper editorials illustrate their coverage of Kevorkian with caricatures of him. Find and bring in a selection of these magazine and newspaper drawings, discussing with your students the primary exaggerated features of each. How do they depict him? As "Dr. Death"? As a professional going about his business with compassion? Do these caricatures of him relate to his essay? Do the cartoons look like a person who could have written this essay?

Suggested Answers for Responding to Reading Questions

1. Kervorkian seeks to show himself as a professional and compassionate person. His language creates a confident tone that is at odds with the magnitude of the controversy of euthanasia. Without showing some doubt, and depicting himself as a lone hero of the sick and dying, he places his credibility in doubt.

2. In paragraphs 12 and 13 Kevorkian compares his specialization to that of a surgeon. The parallels between assisted suicide and surgery are flimsy at best. Kevorkian's patience must give him permission to supply and set up the lethal drugs. This is quite different from a surgeon who acts to correct a medical condition. The plea that he is personally involved with his patients where a surgeon is impersonal is an invalid argument.

3. Kevorkian addresses arguments against his position most directly in paragraphs 12 through 14. Whereas Kevorkian is assured of his ethics, Schneiderman raises many questions and argues that there is much to doubt and discuss as a community. Kevorkian's essay does not refute this position, but merely sets forth an unsubstantiated claim.

Additional Questions for Responding to Reading

1. In paragraph 6 Kevorkian says that known supporters of his work refused to allow him to use their homes. What might their reasons have been?

2. Kevorkian believes that his treatment is more ethical and honest than that of the doctors who perform euthanasia in

hospitals without publicly acknowledging their motives. Do you agree that Kevorkian's approach is more ethical? Explain your answer.

"RUSH TO LETHAL JUDGMENT," STEPHEN L. CARTER

For Openers

1. How does the information about the author influence your view of the essay's claim? What if the author were a doctor, a religious affiliate, or a terminally ill patent? Would those professions bring more or less credibility to the argument?

2. Discuss what are appropriate and not appropriate topics for your class discussion. For example, students may want to share family stories concerning assisted suicides. Consider what you will and will not want to discuss in your class.

Teaching Strategies

1. Different disciplines may study the same situation by asking different questions. For example, a psychologist, poet, anthropologist or a doctor would have different questions and vocabulary for their inquiry and study. Review this essay to show how the questions and discussions are specific to the legal profession of which the author is a member.

2. Does Carter assume that his audience has a basic understanding of the law? Is this a fair assumption? What terms should he have explained or defined?

Collaborative Activity

Divide the class for a mock trial that will decide on the treatment of terminally ill patients should receive. In one case a person with a terminal illness refused life support and died of natural causes. With life support systems, he may have lived a few months longer. In the second case, the person was put on life support and lived for a few months after her diagnosis; however, she was in great pain, could not take care of herself, and felt humiliated by her quality of life. Each group must develop a definition for the right to live and the right to die. Then, in their trials, they must apply their definitions to their debate. The members of other groups will act as jury and supply their verdicts by secret ballot.

Writer's Options

Gordon raises this question, more than once, "Do our mortal lives belong to us alone or do they belong to the communities or families in which we are embedded?" (14). Write an essay that explores the possible answers to that question. Develop an explanation of the assumptions that would underlie possible answers.

Multimedia Resources

1. Ask students to view the movie *Cocoon*. What is movie's attitude toward aging and death?

2. Ask students to read news articles about Kevorkian in *The Detroit Free Press*, *The New York Times*, and newspapers in other countries, such as Canada. Find newspapers with some religious affiliation. Compare language and tone in the different newspapers.

Suggested Answers for Responding to Reading Questions

1. Carter holds that as a society we must take time to consider all the assumptions and implications of this proposed interpretation of the law. Carter warns about hasty judgments or changes.

2. The problem is that many people for whom this right was not intended will claim it. Assisted suicide should be considered on a case by case basis with as little fanfare as possible.

3. Kevorkian would probably applaud the pseudo-logic that results in what seems like a conclusive decision. Carter, however, sees the limitations of legal systems that frequently make judgments too quickly. He argues that by weaving the issue of assisted suicide into the legal system, a profound loss will come with it: loss of lives, loss of respect for life, and loss of a sense of individuals as important members of a community. Carter asks not for a battle of briefs and legal debate, but for moral reflection through popular debate.

Additional Questions for Responding to Reading

1. What is the difference between "constitutional right" and "moral right"? To what extent does law determine moral behaviors in a society?

200

2. What does Carter mean by "moral reflection"? What are some situations where you have resolved a question in this way?

CHAPTER 9

EARTH IN THE BALANCE

Setting Up the Unit: Citing Authorities to Build
an Argument

Each of the readings in this chapter poses an argument. Some are impassioned pleas for help, while others are reasoned calls to action. Regardless of the specific purpose behind the argument, each writer wants to change readers' opinions about or behavior towards some aspect of the environment. Although there is great range in the ways each writer crafts his or her argument, most appeal in some way to a sense of a shared authority. Building on the ideas of others, writers can include information that supports their position. Using others' words, or others' arguments, also allows writers to show that they are not the only ones thinking in a particular way: After all, if several well-known people have argued along a similar line, that is all the more reason why readers should be persuaded. In addition, using others' data demonstrates that the writer has done his or her research and isn't just writing the first thing that comes to mind; thus, it illustrates a commitment to the issue at hand. Finally, it is difficult to argue with a quotation from an "expert," especially if it is appropriately chosen and documented.

Because the environment is a topic that lends itself to scientific explanation and discovery, many of the sources cited in the readings in this chapter are scientists. For example, Rachel Carson is a scientist, and she supplies the reader with ample statistics. Shawn Carlson, and Sally Thane Christensen, who are not scientists, use information from scientific research to strengthen their own cases. Carlson often disagrees with members of the scientific community, which gives his argument a particular strength.

Other writers seek to inspire more political action, and therefore use data collected by governmental agencies and public opinion groups. For example, Al Gore relies on the data generated both by members of the scientific community and by political committees on which he has served. Other writers look to authorities that are less easily quoted but still effective. For example, Chief Seattle appeals to a more spiritual authority. Thus, the type of authority that the writers cite is just as important as the ways in which their information is used. In each case, the writer chooses an authority who, for him or her, represents the community within which the solutions might be found.

You can also use this discussion as an opportunity to explain *plagiarism* to your students. They may be well aware of certain kinds of cheating, such as copying an entire paper, but they may not understand some of the other guidelines for using others as sources. Point out that although the writers of the readings in this chapter do not use MLA or APA documentation, they do make sure that they are giving credit where it is due. Students will need to be reminded that they must always give credit when they are using words — or ideas — that are not their own.

Confronting the Issues

Option 1: Constructing Contexts

The interaction between human beings and the environment can often be as complicated as nature itself. Many environmental hazards would exist without any sort of human intervention; in fact, it is certain that there were many periods of environmental crisis before human beings even populated the earth. But the ways in which human societies have grown, and the ways in which their growth has coincided with attempts to control nature in various ways, has made our relationship with the earth all the more difficult to figure out.

Many believe that the sheer number of humans populating the planet exacerbates problems that would otherwise not be as prevalent. For example, international meetings devoted to the problems caused by overpopulation have shown many of those problems to be environmental. Introduce this concept to your students, and ask them to consider each of the issues in the list below in terms of overpopulation. Finally, if students feel that any particular topic is an issue worth further attention, ask them to create a list of possible solutions. As they read the selections in this chapter, they can then compare these initial plans with the proposals and arguments suggested by the readings.

- Waste disposal
- Petroleum shortages
- Alternative energy sources
- Food shortages
- Famines caused by drought
- Recycling
- Ozone depletion
- Land exhaustion
- Endangered species
- Depletion of the rainforest
- Groundwater quality
- Decreasing biodiversity
- Pollution
- Acid rain
- Environmental racism
- Diseases linked to chemical usage (e.g., Agent Orange)

Option 2: Community Involvement

Some students will be familiar with the book *Fifty Simple Things You Can Do to Save the Earth*. If not, bring a copy to class. For their project, they can write a community-specific version of the book, with resources and reference numbers that match the facilities and needs of your community. To do this, they will need to decide what issues that might matter have "simple" solutions available locally. The readings in this chapter will suggest possibilities. They might also conduct a survey to find out what steps people in your community might be willing to take—after all, what might qualify as "simple" can vary from location to location. Finally, they will need to compile a list of the resources available to people who want to get involved with environmental issues.

Option 3: Cultural Critique

By and large, environmental debates are characterized in the popular media as having two polarized sides, with participants fighting passionately for either economic or environmental concerns. Quite often, issues under discussion have many more than two sides, and more than two groups may have an interest in the outcome of a particular debate. Bring either television newsclips or newspaper articles into class, and analyze them with your students to see how the debate is presented. Discuss why it is expedient to reduce complex issues to two sides.

Option 4: Feature Film

Show your students the 1983 Disney film *Never Cry Wolf.* In it, the protagonist is sent to remote parts of Canada to watch the wolves and see whether they are destroying the endangered herds of caribou. In the process, the main character learns from the wolves, and tries to live as they do. Many of the issues that arise in the readings in this chapter are touched upon in this film, from questions of biodiversity to the inability of most civilians to respect the natural phenomena around them. Ask students to relate relevant scenes in this movie to the selections as they read.

Teaching "Two Perspectives"

Before reading the next two essays, ask students to write about the role they think politics and legislation can play in preserving the environment. To be more specific what can (or should) the American government do to ensure that more care is taken with the environment? The two essays that open this chapter suggest different roles that members of government might play; the first is a letter to a president of the United States, and the other is from a book written by a vice president of the United States. Which piece of writing seems more likely to help environmental causes?

"Letter to President Pierce, 1855," Chief Seattle

For Openers

The final paragraph contains a prophesy, that the "whites, too, shall pass." Discuss this statement with your students. In what sense has this prophesy come true? In what sense did it fail? How likely is this in the future?

Teaching Strategy

Go through the essay with your students, pointing out the different metaphors that are used in conjunction with the white man. What are they? What is the cumulative effect of this imagery?

Collaborative Activity

Split your class into five groups, assigning one paragraph to each group. Ask students to update that paragraph in order to send a message to the current president. What changes should be made? What can stay as written? What needs to be added? Share the new paragraphs with the rest of the class, and revise appropriately to connect the paragraphs into a coherent letter.

Writer's Options

Write a letter to Chief Seattle, explaining to him what the country looks like in the 1990s. Tell him things that he might find heartening, but also let him know the ways in which his 1855 letter has been ignored.

Multimedia Resources

Bring in a copy of the classic public service ad in which a Native American watches cars whizzing by on a superhighway, its passengers throwing trash out their windows. As the ad ends, the camera zooms in on the wizened native American as a single tear runs down his cheek, and a voice-over asks us to stop littering. Show the spot to your students, and ask them to compare the image in the commercial with their image of Chief Seattle. How are native Americans presented? How might this ad be different if it were produced today?

For Openers

In several places in his essay Gore shows how waste disposal is related to race and socioeconomic class. Discuss this link with your class. Do they see evidence of this link in their own lives? Are there controversies about it in your community?

Teaching Strategies

Discuss the following with your students:

1. In the first paragraph, Gore uses a series of words that end with the suffix -less. Why does he do this? What is he trying to tell us?

2. In paragraph 8 Gore compares industrial waste production with our bodily functions. What effect does this analogy have? Does he make the same connection anywhere else in the essay?

3. How does Gore position himself in relation to Congress? What does he accomplish by doing this?

Collaborative Activity

Gore sets up a careful definition of the word *waste* in paragraph 4. In groups, ask students to find examples of each kind of waste that Gore defines throughout the essay. Brainstorm for ways to reduce each type. As a class, discuss the viability of these plans.

Writer's Options

In paragraph 5 Gore writes that each person in the United States "produces more than twice his or her weight in waste ever day." Write about the waste that you think you produce in an average day. What do you throw away regularly? Are there any ways you might reduce your share of the waste? (It might help to look at Lars Eighner's "On Dumpster Diving" in Chapter 7.)

Multimedia Resources

In order to show your students that attention to waste disposal is neither a particularly new nor particularly revolutionary idea, read them passages from Charles Dickens's *Bleak House,* in which Dickens describes the dust heaps that are swallowing up parts of London in the nineteenth century. You might also want to reinforce the reading with the film version of the book, which includes evocative scenes of the garbage mounds.

Suggested Answers for Responding to Reading Questions

1. Chief Seattle says Native Americans feel that land is a family member, a trusted and respected brother. White men, on the other hand, see land as an enemy that needs to be conquered and tamed. Most students will probably agree with Chief Seattle's assessment.

2. Repeatedly calling Native Americans *savages* allows Chief Seattle to subtly, and ironically, point out that the white man's *civilization* is actually much more brutal and savage. He does not intend his characterizations to be taken literally.

3. Answers will vary. Some students will say that reducing and recycling is in fashion, while others might mention the continuation of the Congressional deadlock that Gore describes.

4. Answers will vary.

5. Chief Seattle would be both encouraged and discouraged by Gore's entire book. It would be heartening for Chief Seattle to know that somebody who cared deeply about the environment was in a high governmental position, but he would also be disappointed to hear about the continued decline of the land almost 150 years after his letter was written.

6. Answers will vary.

Using Specific Readings

"THE OBLIGATION TO ENDURE," RACHEL CARSON

For Openers

Ask students to consider whether we have overreacted to the health risks of pesticides and chemicals in the environment. Do they think these substances pose more or less of a risk now than they did in 1962? Likewise, have other issues in the 1990s, such as the depletion of the ozone layer and the destruction of rainforests, supplanted concern about hazardous substances?

Teaching Strategies

You might want to give students additional background on Rachel Carson (notable for being a successful female scientist in the 1960s if for no other reason) and her book *Silent Spring*. When it was published in 1962 *Silent Spring* provoked controversy and caused intense public concern. The *Christian Science Monitor* said, "Miss Carson has undeniably sketched a one-sided picture. But her distortion is akin to that of the painter who exaggerates to focus attention on essentials....It is not the half truth of the propagandist." The *Saturday Review* noted, "It is devastating, heavily documented, relentless attack upon human carelessness, greed, and irresponsibility." In May 1963 President Kennedy's Science Advisory Committee issued a statement in agreement with the basic premise of Carson's book, warned against indiscriminate use of chemicals, and urged more stringent controls and additional research.

Collaborative Activity

Ask groups of students to find out about alternatives to hazardous chemicals and other toxic substances in their local community. Topics might include: agricultural use of pesticides and herbicides, home use of chemicals, production and dumping sites, government regulation of local facilities, and health effects of chemicals used nearby. They can contact a local environmental organization or local agencies and manufacturers. Have students report on how easy—or how difficult—it is to get information.

Writer's Options

Write a letter to Rachel Carson, acknowledging her position but also bringing in some of the realities of your day-to-day life with chemicals.

Multimedia Resources

If you live in or near a farming community, tape some of the fertilizer, pesticide, or seed ads from television and bring them to class. Do these commercials, or the products they advertise, address any of Carson's concerns?

1. According to Carson, we should change the way we live and farm. She also feels there are ways in which we can control pests and weeds naturally, without overusing chemicals, and she believes these alternatives need to be explored further. Responses to the rest of the questions will vary depending on the positions students take.

2. Carson would point out that the chemical war is never won; destructive insects often resurge after spraying, in greater numbers. Chemical control has "only limited success, and also threatens to worsen the very conditions it is intended to curb." Carson says we have to use "knowledge of animal population and their relations to their surroundings" to work toward balance.

3. Answers will vary. She might have described the impact of chemical use on individual people's lives or described specific examples of polluted environments. Perhaps then the sense of danger would hit closer to home.

Additional Questions for Responding to Reading

1. Why does Carson not focus on the negative health effects that have been traced to chemical poisoning? What does she accomplish with this choice?

2. Do you find Carson's statistics shocking? Convincing? Explain your answer.

3. Are you persuaded by the authorities Carson cites? Why or why not?

4. Do we have a "right to know" about chemicals in our environment? And, if we do know, do we have a responsibility to do something about them? Explain your answers.

"RECYCLING: NO PANACEA," WILLIAM RATHJE AND CULLEN MURPHY

For Openers

Discuss the ecological differences between an open system and a closed system. Discuss how the planet earth can be thought of as a closed system. Consider what happens to other closed systems such as backyard ponds, terrariums, and even swimming pools if the ecological balance is destroyed?

Teaching Strategy

In order to propose their claim that the recycling process is not well understood by most consumers, Rathje and Murphy must invalidate the assumptions held by the reader. Ask students to list the evidence used to show the difficulty with recycling. Point out how the examples of recycling practices are ones familiar to the readers. Discuss the importance of choosing evidence with one's audience in mind.

Collaborative Activity

In groups, decide how students can help reduce the garbage production on campus. Consider issues such as glass over plastic cups, computer communications over paper stationery.
Write a proposal that suggests other ways to recycle material and ways to reduce the amount of material for recycling.

Writer's Options

1. Keep a week-long log of your own recycling habits. Observe what you throw away, recycle, or use conservatively. Review your observational notes. What can you claim about your practices? What do you need to know or do to consider yourself a responsible citizen of your environment?

2. Write a process paper explaining how to prepare material for recycling.

Multimedia Resources

1. By keying in "garbage project" as a phrase, one will find different sites related to the authors' work.

2. Visit the following interesting Web page that gives information about projects, employment, and other opportunities in fields centered on recycling http://www.recycle.net. How does this Web site correlate to Rathje and Murphy's ideas?

3. Search your school's home page for sources that promote ecological, social, and economic well being.

4. Browse through the *Garbage Project* web page http://www.arizona.edu~bara/gbg_in~/.htm. Notice that the material is dated. What does this imply?

Suggested Answers for Responding to Reading Questions

1. The authors assume that readers believe that any recycling practices are beneficial to the environment. By challenging assumptions, Rathje and Murphy create the need for their essay. Students may share different stories. Ask them to think about individuals who recycle cans and paper but then are unintentionally wasteful in other ways.

2. The NIMBY factor complicates the ease of recycling. Recycling can be expensive, unsightly, and offensive to the senses. Advantages must outweigh disadvantages for large-scale recycling to win popularity.

3. The loop is closed when used materials are processed to become marketable products. Then the materials have been completely re-cycled. Gathering the used materials is only a part of the cycle.

Additional Questions for Responding to Reading

1. Are attitudes toward recycling specific to certain generations?

2. What do you predict to be the impact of computers and e-mail correspondence on recycling and use of raw materials?

3. What do you consider your impact on your environment?

"THE PLEASURES OF EATING," WENDELL BERRY

For Openers

Consider different blessings said before meals. Bring in several from different cultures. What do the blessings have in common?

Teaching Strategies

1. Consider the negative images and vocabulary Berry uses to describe eating and the food industry. Paragraphs 5, 6, and 7 have particularly good examples. What effect do these negative images have on his argument?

2. Now look at positive descriptions of eating. Study paragraph 21 in particular. How has his tone shifted? Ask students to describe the voice of the author in that paragraph.

Collaborative Activity

Ask students to describe their latest meal in detail. How was it prepared? How long did it take to eat? With whom did you eat? What did you do as you ate? After each member of the group has described that latest meal, have them comment on what Berry would say about their eating practices. Ask them to write a letter in response to Berry and then read the letter out loud to the group.

Writer's Options

Write a rebuttal to Berry's argument. Using a Rogerian approach, first explain Berry's point of view. Then propose your claim, using descriptions and examples as support.

Multimedia Resources

Collect ads for food from magazines, cookbooks, and food boxes. Take notice of the way food is depicted. What is the layout? Look at color combinations and focal points. What was your response when you saw the illustration or photograph? Is any kind of life style suggested by the photography? What is the difference between foods depicted in a cookbook versus depictions in magazines or on boxes?

Suggested Answers for Responding to Reading Questions

1. Berry is quite serious about his terminology. He plays on the word agriculture as "cultural eats." He makes the point that the way we eat reflects on our cultures.

2. Berry writes about all the processing that prepared foods undergo. His descriptions are fairly general, but they emphasize his protest against quick dinners. However, the information provided makes his point clear.

3. Berry suggests that readers grow their own food and visit farms to appreciate the whole process of food production. Food, then, would become more than simply a product to consumers. It would also be a process to admire and respect.

"DEATH IN THE RAIN FORESTS," SHAWN CARLSON

For Openers

Where does Carlson stand? Since he isn't easily characterized as an extremist on either side of the rainforest controversy, ask students to figure out where in the middle he is. This essay presents a good opportunity to show students that most debates are simply not two-sided, but many-sided. What other sides might there be to this one?

Teaching Strategy

Because this is a complex argument, outline its organization with your students. Show them how Carlson moves from one point to the next, building to his conclusion. You might also want to point out the difference between the way he introduces the essay and the way it ends. How does he get from the beginning to the end smoothly? (Point to transitions here and show how important they are in an argument.)

Collaborative Activity

In groups, ask students to take a section of the essay and to list the types of appeals Carlson uses. What type of appeal does he rely on the most? Does he appeal to authority? To reason? To emotion? As a class, tally the results. Are there any surprises? Does he practice what he preaches?

Writer's Options

Carlson writes that most of "the rainforests' biodiversity lies within their bugs" (13). Does this fact alone make you more or less likely to support rainforest conservation efforts? Write about your reaction to the bugs. Does it make a difference in your position?

Multimedia Resources

Bring in any flyers which you (or your students) might have received from environmentally oriented organizations. Analyze them to see if Carlson's criticism is valid. Do they take the approach that he says they do, or are they more reasoned? Is there a difference in tone from group to group? In style? In both? Using his criteria, decide with whose literature Carlson might be most impressed. Which flyers are the most appealing to your students?

1. Carlson is showing that the alarming numbers provided by most environmental researchers are fallacious, figured on an inappropriate model. His entire argument hinges on these faulty calculations, so it is critical for him to explain both the model and its shortcomings clearly.

2. Answers will vary. The tone of the essay is optimistic, despite the urgency of those Carlson is refuting. As he says throughout the essay, the news about the rainforests is bad, but it is not as bad as we have been led to believe. Many conservationists might take issue with Carlson's assertion that the problem has been exaggerated.

3. Carlson's main point, and his main hope, is that reasoned, careful arguments are a better way to preserve the resources we have. Shocking people into action will work in the short term, but it will not effect any sustained or institutional change. If they followed Carlson's advice, organizations would probably not raise very much money at all. With so many issues and groups vying for our attention and our money, Americans have developed a crisis management mentality; in other words, they give to the causes that seem the most urgent.. The panic-inducing approach that Carlson deplores is the most effective one for fund-raising.

Additional Questions for Responding to Reading

1. How many times does Carlson remind readers why he is taking a positive approach to the problem? Do you think he needs to explain it more often? Less often? How does this repetition affect his credibility?

2. In paragraph 17 Carlson addresses some of the other issues that are intertwined with the destruction of the rainforest. What are they? Which of these issues can or should be addressed by the American government? Which are beyond anybody's control?

"IS A TREE WORTH A LIFE?" SALLY THANE CHRISTENSEN

For Openers

Throughout the essay, Christensen states that human life is "the greatest of all natural resources." Given recent attention to overpopulation issues, there are some who might argue that human life is a resource of which we have an overabundance. Tell your students about this position, and discuss the implications of such a debate. Where do they stand?

Teaching Strategy

Discuss Christensen's tone with your students, outlining the shifts in her voice. When does she use humor, and for what purpose? When does she sound angry, and for what purpose? When is she reasoned, and for what purpose?

Collaborative Activity

Christensen mentions irony in several places in her essay. Ask students, working in groups, to identify those examples that Christensen finds ironic. Are they ironic? If so, what is the source of the irony? Each group of students can draft an explanation of the irony (or the lack of it) in Christensen's life and lead a discussion on a particular instance she cites. (You will probably need to define and illustrate the term *irony* for students before discussions begin.)

Writer's Options

Write a response to Christensen, taking on the persona of one of the other writers in this chapter. For example, what would Chief Seattle say to her? What would Barbara Ehrenreich say? What would Jack London say?

Suggested Answers for Responding to Reading Questions

1. Before their value was recognized, yew trees were being destroyed and wasted. Environmentalists might argue that if they had not interceded on behalf of the forest in general, the value of the yew tree never would have been known. The yews could have all been discarded, and no lives would be saved.

2. Answers will vary. Christensen states that it only takes three trees to fully treat one patient, so in the short term many lives could potentially be saved (including her own). On the other hand, she does not address what might happen when the hundred-year-old trees have been completely harvested.

3. The conclusion is effective because it tries to force readers to take a stand. The emotional appeal is strong because she asks readers to include her family in their decision-making process. Answers will vary on the effect of this kind of appeal.

Additional Questions for Responding to Reading

1. This article was written in 1991, and at the time Christensen said that she was "caught in the center of what may become the most significant environmental debate" of our time. Was she right? Are there other environmental debates that seem more contentious? What are they? What is their significance?

2. Does her role as attorney for the United States Forest Service give Christensen more or less credibility? Explain your answer.

3. Did the title of the essay predispose you to identify with this writer? Does it seem misleading?

"IN THE BELLY OF THE BEAST," BARBARA KINGSOLVER

For Openers

Define *mythology* as a system of stories that explains the condition of the world and the way mythic figures who control the world would gain and use their power.

Teaching Strategies

1. List the different mythological names of missiles and other weapons. Discuss the consequences of naming any product after mythical characters.

2. *Personification* ascribes human characteristics to inanimate objects or concepts. Find examples of personification in this article. Discuss the connotations associated with these personifications. What effects do these images have on the argument?

3. Define the term *sarcasm*. Find examples of sarcasm in this essay and discuss effects on the argument. For example, Kingsolver describes one missile as "a huge dumb killer dog waiting for orders"(16).

Collaborative Activities

1. Ask groups of students to collect movie titles from different half-decades (eg. 1995-1990; 1989-1985; etc.). How do the movies from each period depict war? How prevalent is the glorification of war and armaments in each decade?

2. As a group write a poem that begins every other line with the phrase "In times of peace." Create a new myth, referring to present day images and people of peace. The group may have to list possible images first. The group will also have to decide if they will be sarcastic or sincere in their poem. Let them enjoy the creative possibilities.

Writer's Options

Interview teachers or family members who recall the fallout shelters and safety drills of the 1960s. Find out what impact the threat of nuclear war had on them. As you write about the interview, respond to what was said with your own present-day view of nuclear warfare. What might you conclude?

Multimedia Resources

Rent movies such as *War Games* with Matthew Broderick, *The Peace Maker* with George Cloney and Nichole Kidman to show how the threat of nuclear warfare is dramatized. A movie which will generate a lot of discussion, is *The Quiet Earth*. It creates a symbolic rendering of earth after a nuclear disaster.

Suggested Answers for Responding to Reading Questions

1. On the one hand, by associating the missiles with gods, Kingsolver makes the point that the nuclear missiles are treated as if they were part of some harmless fantasy. Fictionalizing the nuclear missiles trivializes the danger of nuclear holocaust. On the other hand, by associating them with ancient Greek gods, Kingsolver shows the incredible destructive power of the missiles. This dual perspective illustrates our culture's ambivalence toward atomic weapons.

2. Opinions will vary on whether her strategy is effective. A mixing of personal narrative with objective information allows the author to give her opinion on the information in a subtle, non-confrontational manner. Eliminating the personal narrative would eliminate the slow unfolding of her perspective. Losing the information would make her narrative merely anecdotal and inconsequential.

3. The poem allows an imaginative view of her perspective and emotional response to the issue. The missiles are compared to

icons of supernatural powers. Answers may vary, but it is
reasonable to agree that her conclusion is effective.

Additional Questions for Responding to Reading

1. What is the effect of putting the reader and narrator in the
    position of child in an adult world? Reference such phrasing
    as "Like compliant children on a field trip, all of us silently
    examined a metal hatch" (19).
2. Note how government authorities and agencies are also
    personified.

Focus: Who Owns the Land?

The following set of essays provides different descriptions and
opinions on human relationships with the environment, particularly
animals. The authors depend on their own experiences with nature
to create an emotional appeal to their readers. Notice that women
write three of the selections. Notice too the use of female perspective
in these essays; question why this perspective is particularly forceful.

"GROWING UP GAME," BRENDA PETERSON

For Openers

Ask students to freewrite for a short period of time about their
associations and definitions of the word *wild*. Discuss what we
admire and fear in the world.

Teaching Strategies

1. Peterson is relating her process of understanding the delicate
    balance between humans and nature. Review the vocabulary
    and choice of incidents that create this vivid idyllic description.
2. Beginning with paragraph 8, the tone of the essay changes to a
    more somber and darker awareness. Note the repetition of the
    word *uneasy*. The word *mortal* at the end of paragraph 9
    punctuates her new awareness.
3. The essay's title suggests a story of growing up. Trace the
    different stages of her growing up with game.

Collaborative Activity

Ask students to explore, in groups, the campus and the surrounding community for examples of animal images. For instance, they could look at architecture, logos, and advertisements for products. What is suggested or gained by these associations with animals?

Writer's Options

Compare Wendell Berry's attitude to food with Peterson's. What is each assuming about human's relationship to the earth? Choose a format that would best explain their positions. Students could write poems, plays, essays, or letters to the authors. Ask students to be sure to identify themselves in their writing, noting their own connection to the subject of food.

Multimedia Resources

The movie *City Slickers* with Billy Crystal is about city-dwellers' encounter with wildness, nature, and their place in the world. Ask students to view this film, and then review together the scene where Crystal's character aids in the birth of the calf. Why does he take this calf home as a pet? Would Peterson like this movie?

Suggested Answers for Responding to Reading Questions

1. Peterson's purpose is to relate her process of understanding the delicate balance between hunter and game and coming to terms with that balance. She tells about her childhood and talks about her discomfort in order to make herself credible to the reader.

2. Paragraph 7 shows her coming of age. She is allowed to go out with the adults to be part of the hunting expedition, not as a hunter, but as the receiver of the game of the day. She shows the expedition as pleasurable and describes the reverence of the hunters for nature.

3. Readers cannot understand the full impact of what the phrase "horror and awe and kinship" means to Peterson until the last paragraph. She admires the food chain, and thus acknowledges her own mortality. Ehrenreich would agree that humans are very much like animals. However, her view of humans is not as complimentary as Peterson's.

Additional Questions for Responding to Reading

1. Have you ever eaten something that you enjoyed until you discovered what it was that you ate? What are the origins of your dislike for certain new foods. What is Peterson's response to eating domestic meats?

2. Highlight adjectives and adverbs in this essay to appreciate the number of descriptive words used in this essay.

"OUR ANIMAL RITES," ANNA QUINDLEN

For Openers

Tell students about the controversy in Colorado some years back, when a family built a house in the mountains and their child was attacked by one of the indigenous mountain lions. What would Quindlen say to these parents? What would your students say? Should the parents be able to press for legislation, or extermination of these predators, in order to assure safety for their family?

Teaching Strategy

Quindlen spices her writing with asides and quick transitions. Identify these strategies and discuss them with your students. For example, what is the effect of Quindlen's comment "All you who lost interest in the dog after the baby was born, you know who you are"(7)? Are these strategies that you would encourage your students to adopt? Discuss your reasons with them.

Collaborative Activity

In paragraph 6 Quindlen notes the irony of calling people "animals" when they behave in horrible ways that would never occur to actual animals. In groups, brainstorm for lessons that we can take for ourselves from the real behavior of animals. Each group can pick one species to analyze, and generate a list of the qualities that humans should consider nurturing themselves. As a class, discuss the traits that are common to all the lists.

Writer's Options

Quindlen writes paragraph 3 from the perspective of the bear. Do this for an animal which you regularly have contact; it could be a pet, or a squirrel that you walk by each morning, or the fly on your window. What might that animal be thinking about you and the way you live?

Multimedia Resources

Bring in an episode of the television cartoon *Yogi Bear*. Ask students to describe the ways in which humans are depicted in this cartoon. How about the different animals? Does the cartoon seem to reinforce Quindlen's argument, or does it have a different message? Do the two mediums share any of the same techniques?

Suggested Answers for Responding to Reading Questions

1. Answers will vary. Certainly much of our history presents stories that illustrate our sense of our own superiority over other forms of life (see Barbara Ehrenreich's "The Myth of Man as Hunter" in this chapter). Likewise, we tend to think that a human life is the most valuable resource on earth (see Sally Thane Christensen's "Is a Tree Worth a Life?" also in this chapter).

2. Answers will vary. Quindlen ascribes very human characteristics to the animals that she is describing, which could be considered both a form of speciesism and proof of her overromanticizing. Students will have differing opinions, but the examples of the deer following the same trails and the turtles biting the swimmers evoke a sympathy for the animals and a willingness to ascribe insight and dignity to them.

3. Like William Stafford, Quindlen is haunted by her encounter with an animal. However, Quindlen's bear continues to live and enjoy his freedom. Stafford's deer was vanquished by encroaching civilization.

Additional Questions for Responding to Reading

1. When Quindlen first sees the bear, she thinks of him in relation to a television show. What comment is she making by opening her essay this way?

2. Compare Quindlen's approach to animals with Jane Goodall's in "A Plea for the Chimps" (Chapter 8). What do they have in common? Where might they disagree?

"THE MYTH OF MAN AS HUNTER," BARBARA EHRENREICH

For Openers

At the end of paragraph 8 Ehrenreich writes that "the earth's scariest predator has been ourselves." Discuss the implications of this statement with your students. What does she mean? Can they provide some examples to illustrate her point?

Teaching Strategy

Ehrenreich takes aim at the anthropological community in paragraph 3, where she describes their explanations of the puncture marks in a fossilized skull. Use this as an opportunity to discuss "guiding assumptions" to your students. In other words, how do the assumptions that we make about the way the world works, and our role in the world, guide our conclusions?

Collaborative Activity

In groups, ask students to concoct a myth that might replace the "man as hunter." What other metaphors would work to show humanity's relationship to the rest of the world? What implications do the new metaphors have for the ways in which we interact with the environment? Ask each group to present its metaphor, along with the implications it carries. Vote on the one that works best.

Writer's Options

Write about a fight that you had with nature—and lost. What happened? Was any part of the outcome within your control? What did "losing" mean in this context? Is it something that you are afraid will happen again? What would you do if it did?

Multimedia Resources

Bring in some old textbooks that talk about the earliest human beings. How do these texts describe the ways in which people interacted with the environment? Are human beings really depicted as hunters and gatherers, or as something else? How much of the myth that Ehrenreich describes seems to be presented in these types of textbooks? Do they mention any research supporting other views?

Suggested Answers for Responding to Reading Questions

1. Most likely, Ehrenreich's view on sport hunting is that it is unnecessary and self-congratulatory. Since its purpose is primarily entertainment, not survival, it merely reinforces the sense that "historically speaking, it was our side that won" (7). Sport hunting, in this view, is the myth enacted on a lesser and insignificant scale.

2. Ehrenreich suggests that the myth endures because we want it to, even need it to. (Similarly, when we go to horror movies, we want a happy ending, and we want to feel secure that we can overcome our predators.) In other words, the myth gives us constant reassurance of our superiority.

3. Each author thinks that humans share animal traits in their behavior and life cycles. Peterson says we are the same under our skins. Peterson is disturbed by hunting, but finds an honorable act in hunting. However, Ehrenreich sees very negative aspects in humans because of their selfish aggression and their inability to always protect themselves from powerful natural forces. She sees hunting as selfish arrogance. Quindlen does not think that humans should hunt game or take away their homes in order to build their own.

Additional Questions for Responding to Reading

1. The title "The Myth of Man as Hunter" seems as if it could refer to either men alone, or to humanity as a whole. Which do you think Ehrenreich intends? Why do you think she leaves it ambiguous?

2. How do you think Ehrenreich would explain the growing popularity of laser tag games? What other games allow groups to practice or vent their violent behavior?

"Traveling Through the Dark," William Stafford

For Openers

Ask students whether they have ever seen a dead animal lying beside the road. How did their reactions differ from those described in this poem? How do they account for any differences? Do they think that most people react more emotionally to finding a deer (or a pet) than to finding another animal—say, a rabbit, squirrel, or possum? Why?

Teaching Strategies

Discuss the following with your students:

1. Stafford made the following comments about his own work: "when you make a poem you merely speak or write the language of every day, capturing as many bonuses as possible and economizing on the losses; that is, you come awake to what always goes on in language, and you use it to the limit of your ability and your power of attention at the moment." Does his work seem to reflect this attitude?

2. Contrast the images of nature and the images of encroaching civilization. For example, the road is bringing traffic to the wooded mountain and the car purrs. How is the speaker of the poem caught between the wilderness and its beauty and his own practical view of the situation he has stumbled upon?

Collaborative Activity

Split the class into four groups, assigning to each group one of the first four stanzas. Each group should then consider the theme of the collision between nature and human civilization; the interrelatedness of that relationship; and the speaker's identification with nature and with civilization. Once they have formulated a response, groups can report to the class and consider the final stanza in light of the insights drawn from their group conversations.

Writer's Options

Write about a time when you observed a conflict between nature and civilization. How was it resolved? Why has it stayed with you?

Multimedia Resources

Many commercials advertising the safety of their anti-lock brake systems will show a car, often driving at night, stopping suddenly to avoid hitting a child or an animal. Show one or two of these ads to your students. Why is this such a popular theme in car ads? Does it demonstrate something else besides the efficacy of the new braking feature? What sorts of people are shown driving these cars at night? Finally, remind students that most driver education classes admonish people not to slam on the brakes to avoid hitting animals. What do they think of this advice?

Suggested Answers for Responding to Reading Questions

1. This line tells readers that the driver has been in this type of situation before; it also, in a matter-of-fact tone, articulates the prevailing attitude that dead animals are not only hazards, but also unpleasant and therefore easier to ignore when they are out of sight. The line makes the speaker's initial hesitation a bit difficult to understand, and it helps to explain his final action.

2. "Beside that mountain road I hesitated," "I could hear the wilderness listen," and "I thought hard for us all" suggest the speaker's ambivalence about the natural world and his awareness of his separateness from it. He pauses, hesitates, hears, thinks hard — he is attentive and respectful — but, still not one with nature; he must act as a human being.

3. If the speaker feels only "horror and awe," he does so only momentarily. Ultimately pragmatism takes over. He hesitates for just a moment and then pushes the deer into the river.

Additional Questions for Responding to Reading

1. Why do you think Stafford titled his poem "Traveling through the Dark"?

2. What would you have done in the speaker's place? Would you have stopped the car in the first place? Would you have tried to find some way to save the fawn?

CHAPTER 10

MAKING CHOICES

Setting Up the Unit: Using Emotional Appeals to Inspire Commitment

The readings throughout this chapter are designed to compel readers to think more seriously about the choices they make and, in particular, to encourage them to consider making choices that might not be more thoughtful than popular. These writers are hoping to get a commitment from the reader: a commitment to choose, to question, to act—and, maybe, to resist acting in certain prescribed ways. In order to get this commitment from their audience, these writers try to get their readers emotionally involved in their discussions. They have a number of strategies for doing this, often called *emotional appeals*, or *pathetic appeals* (named after the Greek word *pathos*, not because they are deemed ineffective; on the contrary, they can be very effective indeed).

One strategy is to build a relationship with readers, perhaps by referring to them, by addressing them directly, or by inviting them to participate in the creation of the text itself. For example, Martin Luther King, Jr., writes his letter directly to a particular group of people, and he refers to them—and their shared backgrounds— throughout the letter.

Humor is another technique writers can use to engage readers. For example, George Orwell pokes fun at himself with his constant introspection. Humor can act like a hot pad, allowing the reader and writer to handle serious topics. Humor binds the reader to the writer because in accepting the humorous view of the situation a reader often accepts criticism or a new point of view.

Most often, emotional appeals are discussed in terms of specific word choices. Often writers will use language with emotional connotations in order to elicit sympathy from readers. For example, both Nicholas Jenkins and Garrett Hardin strengthen their arguments by using words that can be considered loaded. They use images that paint a vivid picture for readers, making it impossible for them to ignore the writer's position. Annie Dillard's initial description of the deer causes the same emotional reaction, even though she herself is not outwardly disturbed by the deer's predicament.

Not all of the language is emotional, however. Some writers, like Deborah Lipstadt and Stanley Milgram, use rational, even scientific, rhetorical strategies to create an emotional response in the reader. They rely on the impact of their exhaustive work to overwhelm and move their readers. Sheer depth of analysis and repetition of examples is enough to compel readers to believe — and to care.

Encourage your students to try some of these techniques in their own writing. Of course, you will need to remember that when they overdo these methods, they run the risk of alienating readers who aren't already sympathetic. (You might cite John Leonard's "Why Blame TV?" (Chapter 4) or Judy Brady's "Why I Want a Wife" (Chapter 5) as examples of writing that goes too far. Still, emotional appeals are a strategy that students might want to have at their disposal.

Confronting the Issues

Option 1: Constructing Contexts

In "Lifeboat Ethics: The Case Against 'Aid' that Harms," included in this chapter, Garrett Hardin argues that the world's limited resources cannot be distributed to everyone who needs them. He uses the lifeboat metaphor to help make his case: If we let everyone climb aboard, we all will drown. A few can safely be allowed into the boat, but the question of how those few will be chosen has no simple answer.

Ask your students to imagine they are on a real lifeboat after a disastrous sea accident. The passengers can allow only one person to join them without risking their own safety. Those in need of rescue include the following:

- Bill McKay, twenty-seven. Outstanding African-American athlete; last year's top NBA pick. Popular speaker at inner-city high schools where he urges students to stay in school.

- Jenny Lang, twenty-five. High-school dropout and single mother of four preschool children. She has no other family and depends on welfare for financial support.

- Frank Warfield, sixty-five. Philanthropist and Holocaust survivor who has been a generous contributor to numerous charities. Father of six, grandfather of fifteen. Has a heart condition.

- Jane Gordon, forty. Lawyer and advocate for the homeless and for battered wives. Volunteer literacy tutor, Special Olympics coordinator, and soup kitchen worker. She is the local leader for gay rights.

- Freddy Logan, twenty-two. Former gang member and cocaine addict, now employed and a part-time college student. Sole support of this fourteen-year-old brother, who looks up to him as a role model.

- Ana Garcia, thirty-five. Medical student. Single, no children. Recipient of a special community scholarship for Hispanic-Americans, she is a straight-A student who plans to practice medicine in an impoverished area. She is the first person in her family to have graduated from college.

- John Ruscomb Tucker, eighteen. First-year student at Yale, where he plans to major in business. Only son of a U. S. senator who is often mentioned as a future presidential candidate.

- Elizabeth Pfam, fifteen. Daughter of a Vietnamese mother and an American GI father she never knew. Talented musician; enrolled in a program for intellectually gifted students. Came to the United States as a small child. Her mother, who speaks little English, relies on Elizabeth to help her cope with American society.

- Justin Moore, thirty-six. Newly ordained minister, who has been assigned to a parish in an impoverished part of a large urban area. Directs and performs in community theatre, organizes work programs to renovate and repair homes for the poor, leads a food delivery service for shut-ins. Is currently nursing his lover through the final stages of AIDS.

- Joanne Templeton, thirty-five. Homemaker, Little League coach, Cub Scout den mother, school board member. Actively involved in reviewing grade-school textbooks, rock lyrics and videos, and films to make certain content is not at odds with her strong

Christian values. Has five children, one of whom has Down's syndrome.

Without discussion, ask students to designate their choices by secret ballot; then count the votes and put the results on the board.

Discuss the candidates, beginning with the one who received the most votes and working down to the one deemed least worthy of rescue. Ask your students to mention reasons why the person should be saved, as well as arguments against that individual. As your students discuss the candidates and try to reach consensus about the top choice, they will be forced to make moral judgments about the value to a society of money, education, youth, family, and talent. They will also have to consider factors that make a person seem undeserving such as health problems, antisocial habits, or an unconventional lifestyle. Because students are likely to have varying ideas about the relative value of different individuals to society, they will need to defend – and perhaps to reconsider – their choices.

At the end of the class period, take another vote, again by secret ballot. Ask each student to indicate whether his or her vote has changed. If the discussion has been serious and honest, some votes will certainly be different.

Option 2: Community Involvement

Many of the questions raised by this chapter revolve around the issue of individual responsibility to a group. Ask students to discuss what their society offers them, encouraging them to be as specific as possible. In what ways can students give something back? Their project for this paper can be to find the outlets for contributing to their community in ways that are appropriate to their talents and desires. Ask students to explore opportunities for community service projects through their own school or department. As a part of this process, students may need to discover that there is "something in it for them" when they participate in a positive way. The resulting project can be a report or proposal outlining the ways in which this specific group of students contributed and the ways in which their contribution benefited them.

Option 3: Cultural Critique

Many of the readings in this chapter argue – in one way or another – that individuals should take responsibility for their own choices and for the consequences of their actions. The focus on the *individual* is paramount; most of the writers are encouraging readers to do more than simply follow the latest wave, unhesitatingly and uncritically. Quite often we see these individuals who stand against mainstream as heroes, given special recognition for their strength (for more on heroes, see Jenny Lyn Bader's "Larger Than Life" and Arthur Levine's "The Making of a Generation" in Chapter 7). Ask students to bring in images of "against the grain" heroes: pictures, articles, television specials, songs, posters, and so on. Analyze with them how many are presented alone, and how many appear as part of a group or coalition. How many of the images depict a renegade acting alone, and how many depict the strength of communities and cooperation? Discuss this difference with your students. How much power does an individual have when acting alone? When acting with others for a shared goal? Which method is more effective for long-term social change?

Option 4: Feature Film

To illustrate the difference between social responsibility and uncritical obedience to authority, arrange for a screening of Steven Spielberg's film *Schindler's List*. This movie ties together many of the strands that intersect in this chapter, from humanitarian action to civil disobedience to questioning authority to individual social action to issues involving the Holocaust. The film is long, but well worth viewing for the discussions that it might generate. Encourage students to approach the film as a series of *choices* that individuals make.

Teaching "Two Perspectives"

These two poems relate very simple situations: a walk in the woods and a child's birthday party. Yet these disarmingly simple poems dramatize the choices we make to be individuals in a possibly indiscriminate society. Ask students to consider how purposefully they choose to take action and how easy it is to be distracted. How often do we react thoughtlessly and how easy is it to arrive at a place we never intended. Consider social, academic, and personal examples.

"THE ROAD NOT TAKEN," ROBERT FROST

For Openers:

What exactly is "the road less traveled by"? Give some concrete examples of such roads in our lives. What well-known individuals have taken such roads?

Teaching Strategies

Discuss the following questions with your students:

1. During which season does the journey take place? Why does the season seem appropriate to the poem?
2. Chart the poem's rhyme scheme and meter. How are they consistent with the poem's subject and theme?

Collaborative Activity

In groups, ask students to discuss whether the poem is really about the road the speaker chose, which "made all the difference," or, as the title suggests, about the road *not* taken. Each group can present their findings to the class, and a full-class discussion can proceed from there.

Writer's Options

What two roads diverged in your life? Which path did you take? Why? Was it the right choice? How do you know? Describe you decision and its consequences.

"ETHICS," LINDA PASTAN

For Openers

Ask students to define ethics. Bring your college catalog (several if possible) to class so you can search your college catalog descriptions for courses on ethics. Which disciplines teach courses related to this concept? What is taught and studied in these courses?

Teaching Strategies

1. Discuss why this poem is considered poetry and not prose. Written as prose, it seems like a reflection or a journal entry. Discuss the special language of poetry.

2. Notice the many references to time from "so many years ago" to "This fall." What effect do these references have on the reader? Discuss how repeated elements create a total emotional effect.

3. Discuss the relationship between art and life. How does one support and intensify the other?

Collaborative Activity

Ask students, in groups, to recall questions and debates posed to them as young students that at that time seemed to have certain answers, but that now seem quite complex.

Writer's Options

Write an essay that argues the value of art museums to a community and to a culture. Consider the financial commitment and support an art museum requires. Would a community choose more wisely by supporting student scholarships, soup kitchens, and the homeless instead of spending money for housing and promoting art work?

Multimedia Resources

Bring in several Rembrandt prints in textbooks or slides from your school archive. Allow students time to look closely at the paintings and to write reflective responses to the art work.

Two Perspectives: Suggested Answers for Responding to Reading Questions

1. He chooses the "road less taken" for its mystery. Perhaps the speaker is looking for an alternate way of living.

2. The paths suggest more than a way through a wood. The path becomes a metaphor for a life's journey.

3. The speaker celebrates the satisfaction in taking a risk by taking a different road. Nothing will make a difference for Pastan's speaker because the passage of time cannot be altered. What could make a difference was not seeing the problem as a dilemma. After all, the old woman could have grabbed the painting and saved both herself and Rembrandt!

4. The speaker is asked to choose between art and an individual life. She is asked to make a value judgment. The teacher offers no answers, but rather poses the dilemma. Most likely, she

expects the students to value the life of the woman over the art work.

5. Answers may vary. The speaker in Pastan's poem seeks a new question, one more relevant to her way of thinking. Frost's speaker must choose because the path is his own, and he cannot stop moving forwards.

6. As life interprets art, and art interprets life, the woman and the painting are one. The season of autumn is represented in the painting and is also the natural image of the woman's stage in her life. The teacher's question has no satisfying answers. Rather, the question should lead to discussion and debate of ethics and social values.

Using Specific Readings

"THE DEER AT PROVIDENCIA," ANNIE DILLARD

For Openers

Ask students to consider the phrase "These are not issues; they are mysteries" (14) as it applies to the events described in the essay.

Teaching Strategy

Have your students look for images of isolation, of alienation, and of being an outsider. These images might serve to support Dillard's sense that she is unable to intervene in the suffering and distress of another being.

Collaborative Activity

Ask students, working in groups, to list events that they consider to have been "unfair." In class discussion, try to agree on a definition of *fairness*.. Would Alan McDonald's wife agree with the class's definition?

Writer's Options

Write about a time when you witnessed a person or animal in need or in pain and were unable to intervene. How did you respond? What caused you to respond the way you did? Has the event haunted you? What is its significance to you?

Multimedia Resources

Even if you have already assigned it, read aloud William Stafford's short poem, "Traveling Through the Dark" (Chapter 9). How are Stafford's and Dillard's views alike? How might they disagree?

Suggested Answers for Responding to Reading Questions

1. Dillard could have insisted that someone put the deer out of its misery, but she chose not to interfer with the local custom. She was a visitor and an observer. Dillard seems to have accepted the notion that pain is inevitable and that the world is not always ordered or just. Her final utterance in Spanish underscores her sense of impotence.

2. For Dillard the paradox of hating violence yet eating meat is no different from any other mystery in the world. It cannot be explained in neat, reasonable terms.

3. Both man and animal are "victims" in an often cruelly illogical universe. Therefore, there is a reasonable connection between the two situations.

Additional Questions for Responding to Reading

1. Do you think the act of writing this essay might have served any psychological or emotional purpose for the author?

2. How do you account for Dillard's *not* having left the village at the sight of the suffering deer? Does she lose your respect by staying to watch?

3. What could Dillard have said to Pepe at the end of the essay to better express her feelings about the deer?

4. What assumptions about gender are present in this essay? What is assumed by Dillard? What is assumed by the men with whom she travels?

"SHOOTING AN ELEPHANT," GEORGE ORWELL

For Openers

Consider the tension between power and powerlessness in the essay. Who (or what) seems to be in control?

Teaching Strategies

Give your students some additional background on colonialism, and on Orwell. Always an aggressive spokesperson for the poor, Orwell challenged the political orthodoxy, opposed imperialism and aristocratic privilege, and took the stand for English popular culture of the 1940s. He perceived the artist's role as an important one: Art, he believed, served to extend human sympathies, and the artist provided an intellectual commentary and helped to shape society. Orwell himself did this, particularly in relation to British colonialism, showing how otherwise kind and moral individuals participated in and thus helped to perpetuate a cruel and exploitative system.

Collaborative Activity

Arrange a series of debates based on Orwell's dilemma. Have teams of students craft arguments that (1) the first responsibility of any authority figure is to fulfill the duties required of a leader, or (2) an individual's primary obligation is to his or her own sense of justice and morality.

Writer's Options

Have you, like Orwell, ever been in a situation in which you wanted to act one way but were forced by circumstances to act in another? Write about this incident. In retrospect, how might you have resisted?

Multimedia Resources

1. To show a film that was produced within years of Orwell's writing this essay, show students parts of the 1939 film *Gunga Din*. It will provide them with a visual sense of British colonialism as seen through the eyes of other officers. Does the narrator of Orwell's essay seem like a character in the movie? Which one?

2. An old map, showing all England's turf, would also be a terrific eye-opener for most students, who won't know or remember when the "sun never set on the British empire."

Suggested Answers for Responding to Reading Questions

1. Answers will vary. Although Orwell seems to imply that he feels some degree of self-recrimination for shooting the elephant unnecessarily, it could be argued that he had little choice. His will to act might have been among the greatest victims of imperialism.

2. Answers will vary

3. Although the essay seems to be primarily directed at self-examination, Orwell's entrapment in the web of authority imposed by the imperialist system seems to be his most pressing concern. The essay does both.

Additional Questions for Responding to Reading

1. What is the tone of the essay? What would you judge to be the narrator's attitude toward his younger self and his past behavior?

2. Is the narrator more critical of himself, or of the system within which he is living and working? Explain your answer.

3. How would you define what Orwell calls a "roundabout" way of coming to enlightenment? What is the enlightenment in this essay?

4. Explain how Orwell's dilemma is both like and unlike Annie Dillard's in "The Deer at Providencia" (p.724). Are the two writers dealing with different issues?

"CIVIL DISOBEDIENCE," HENRY DAVID THOREAU

For Openers

Thoreau writes, "How does it become a man to behave toward this American government to-day? I answer, that he cannot without disgrace be associated with it." Ask students if they think Thoreau is being unpatriotic. Does his statement, made over one hundred years ago, seem dated? How might this statement apply today?

Teaching Strategies

Discuss the following with your students:

1. Provide some additional background on this essay. Thoreau's residence at Walden Pond was disturbed for one day when he was imprisoned for having refused to pay a poll tax to the government in support of the Mexican War. The essay, which outlines his belief in passive resistance, was originally delivered as a lecture. This influential argument in support of the individual's obligation to his or her own highest standard of ethics was later printed in Elizabeth Peabody's *Aesthetic Papers* (1849).

2. Explain the key concepts of transcendentalism to your students.

3. Thoreau's essay might serve as a model for various rhetorical devices. Look, for example, at his use of *paradox, periodic sentences* (those in which the main idea is postponed until the end), and *rhetorical questions*. Encourage students to identify and replicate these and other strategies.

Collaborative Activity

Ask students, working in groups, to analyze sections of the essay at the paragraph level. Can they isolate and then paraphrase each topic sentence? How do transitions help connect sentences and paragraphs?

Writer's Options

1. Do you agree that only "a very few" "heroes, patriots, martyrs, reformers, in the great sense" have served their country with their consciences? And that they have been treated as enemies? Write about one such figure who is particularly meaningful to you.

2. Thoreau suggests that his night in jail prompted him to view familiar things from an entirely new perspective. Describe an experience that made you see things differently.

Multimedia Resources

Show students scenes from the movie *Gandhi* in which the hero articulates his philosophy of civil disobedience. What parts sound as if Thoreau might have written them? Which parts are unique to Gandhi's particular context?

Suggested Answers for Responding to Reading Questions

1. All the phrases, aimed ostensibly at the American political structure, reflect Thoreau's sense that the individual's obligation is to the highest form of morality, a morality that transcends government.

    a. "That government is best...." implies a political choice in favor of the right and competency of the individual to serve as the highest possible authority in the government.

    b. "All men recognize..." a government is subject to its people, therefore, when individuals are in conflict with their government, they must obey their own sense of moral obligation. The phrase implies both a moral and a political choice.

    c. "All voting..." shows Thoreau's belief that his fellow citizens do not take their right to vote seriously enough, that they do not see each vote as a vital moral decision.

    d. "Under government..." implies that our failure to act against injustice incriminates us. If political injustice exists at all in this country, the only way for the truly moral individual to behave is to demonstrate opposition and to face
    political imprisonment for a moral issue.

    e. "I did not see why..." shows that the inequities of the tax system are both a political and a moral issue. The paying of taxes is not simply a passive duty that we perform in order to avoid breaking the law. Thoreau urges us to comprehend the real consequences of our actions To whom are we giving that support, and for what cause?

2. Answers will vary.

3. Answers will vary.

Additional Questions for Responding to Reading

1. What do you understand Thoreau to mean when he calls for a "man who is a man"? Why does he say there are so few "men" among Americans? Can he be accused of sexism?

2. Which of Thoreau's statements seem to you particularly surprising? Would you disagree with his advocacy of any specific actions? How does he reverse your expectations of what constitutes lawful behavior?

3. What is the dominant tone expressed in the essay? Is Thoreau primarily critical of America, or does the essay suggest a wider focus?

4. Thoreau's ideas influenced Gandhi, Martin Luther King, Jr., and Nelson Mandela, among others. What specific ideas do you think appealed to each of these individuals? Which of Thoreau's ideas might contemporary advocates of political reform reject?

5. Thoreau asserts that when an individual acts on his own principles, the far-reaching gesture divides states and churches, families, and even the individual from him- or herself. What does this mean? Do you see examples of this division occurring in society today? If so, how does it manifest itself?

"LETTER FROM BIRMINGHAM JAIL," MARTIN LUTHER KING, JR.

For Openers

Ask student volunteers to read paragraph 14 aloud. What rhetorical and stylistic strategies do students "hear"? Why are they effective?

Teaching Strategies

Discuss the following with your students:

1. Provide some additional background on this letter. In his book *Why We Can't Wait*, which includes "Letter from Birmingham Jail," King describes the process of training recruits in nonviolent techniques. He discusses, too, his unusual decision to use children in a demonstration, in Birmingham. Photographs taken during this demonstration showing children as well as adults being beaten down by jets of water and attacked by police dogs, provoked outrage throughout the nation. As a result of this exposure, President Kennedy sent Justice Department negotiators to Birmingham, and a settlement was effected within one week. In this letter,

addressed to his fellow clergy members and written while he was incarcerated in the Birmingham jail, King demonstrates his power as a speaker, writer, and leader. Clearly, King had taken to heart Thoreau's message: "One has a moral responsibility," writes King, "to disobey unjust laws."

2. Discuss this letter in conjunction with Thoreau's "Civil Disobedience." Ask students what King seems to have learned from Thoreau. Does King carry Thoreau's points to an extreme? Do you imagine Thoreau would have supported King's movement?

Collaborative Activity

Ask students, working in groups, to identify passages from the essay in which King moves from a general discussion of the movement to a specific recounting of personal details. Ask whether the shift into a personal tone is effective. What is its impact?

Writer's Options

Write a newspaper editorial in response to King's position. How might it be viewed today? What background might you need to offer readers? What applications do you see to today's society? If possible, relate your position to a current event in your community, or in the national or world news.

Multimedia Resources

1. If you want to show some contrast to King's position within the African-American community, show some of the later scenes from Spike Lee's film *Malcolm X*. Contrast Malcolm X's position, "by any means necessary," with King's nonviolent resistance.

2. To show additional footage of the Civil Rights movement, bring in segments of the PBS series *Eyes on the Prize*. Find out from your students which parts of the movement are most intriguing or confusing to them, and show scenes that correspond with your students' interests. We recommend those scenes discussing Birmingham, Selma (Rosa Parks), and Little Rock.

Suggested Answers for Responding to Reading Questions

1. The clergymen have admonished King for being what they consider "extremist" and impatient. King argues that time must be used constructively because oppressed people eventually fight back. Nonviolent action offers the only alternative to violence because it offers a creative outlet for discontent.

2. Answers will vary. Some people might argue that the collapse of the former Soviet Union fulfills King's prophetic statement, and others might point to advances in South Africa. Those who disagree with King may point to situations in Bosnia, Haiti, and urban areas in the United States.

3. King was a powerful and moving preacher and orator, and his writing style utilizes his strengths as a speaker. He speaks to the clergy as a fellow member, using strategies that will be familiar and comfortable to them. Likewise, he continually refers to Biblical passages, showing his credibility as a minister and strengthening his bond with his audience. Individual students might find this style distracting, while others will feel it is appropriate and effective in showing that King was a compassionate, vital, and highly intelligent person.

Additional Questions for Responding to Reading

1. King writes that the South has been bogged down in a "tragic effort to live in monologue rather than dialogue." What does he mean? In what ways does his "letter" help to establish a dialogue?

2. What steps are taken to prepare for direct action? Do you understand the necessity for each of these steps? Explain your answer.

3. Who are the people who are affected by segregation, according to King? Do you agree with his assessment? Do all parties suffer to the same extent?

"LIFEBOAT ETHICS: THE CASE AGAINST 'AID' THAT HARMS," GARRETT HARDIN

For Openers

Can Hardin's argument be applied to the sick? The elderly? The disabled? What dangers face a society that gives no aid to its weakest citizens?

Teaching Strategy

Some students might find Hardin's attitude cold and arrogant. Rather than focusing at once on the essay's emotionally charged issues, you might begin by examining the logic with which Hardin presents his argument. Analyze the organizational structure of the essay.

Collaborative Activity

If you have not done the *Confronting the Issues* exercise on page 223 with your class, do so now. Students can work in groups to decide who should be allowed into the boat. As a class, tally the responses of each group. Who wins? Who loses? Why? What do the results tell you about who is valued in our society?

Writer's Options

1. Write about a time you were caught in the tension between following your conscience ("doing the right thing") and struggling for your own survival. Which side won? Was any compromise possible?

2. Write a letter to your U.S. senator arguing for or against sending aid to Somalia, Ethiopia, or Haiti. As you make your case, support or refute Hardin's arguments, perhaps citing him as you do so.

Multimedia Resources

Now something of a cult classic, the 1973 film *Soylent Green* depicts a futuristic vision of what might happen if Hardin is right. It shows an overpopulated and hungry New York City in the twenty-first century, a world in which people have no other options but to feast on their own dead—regulated, packaged, and produced through government agencies. The best scene to show your students would probably be the one in which Edward G. Robinson reminisces about the good old days of "real food" in the twentieth century.

Suggested Answers for Responding to Reading Questions

1. Answers will vary. Hardin presents a complex argument as if it has only two possible alternatives. One option he fails to account for is that wealthy nations can offer forms of assistance other than increased technology. Not only do wealthy nations help poorer nations increase food productivity, but they also help provide education in medicine and family planning.

2. Answers will vary. The lifeboat image makes Hardin's reasoning clear, but it is not equally relevant to all his arguments. The metaphor of the commons is more appropriate to Hardin's argument about the World Food Bank, since this image suggests that there are negative consequences to making resources widely available.

3. Answers will vary. Hardin clearly identifies himself as a member of a wealthy nation; students might argue that his attitude would be different if he were one of the swimmers outside the boat.

Additional Questions for Responding to Reading

1. What does the environmentalists' metaphor of the spaceship suggest? How do the ethics of the spaceship differ from those of the lifeboat?

2. How are liberal thinkers characterized in this essay? Do you agree with Hardin's portrayal?

3. What would seem to be Hardin's basic view of human nature? Is it at all hopeful? Explain your answer.

"DOG LAB," CLAIRE MCCARTHY

For Openers

1. What are professors' responsibilities toward their students? Are the professors in this essay acting responsibly?

2. What responsibilities do students have for their own education? Ask students to look through the university material that they received at orientation or in the residence hall or the information in the college catalog. How is student responsibility described in these publications?

3. Some educators say that college work permanently changes a student's way of thinking. Do you agree?

Teaching Strategy

1. Review how coherence is achieved through repetition of key words and parallel structure. Explore the essay's effective use of parallel structure. Show how parallel structure is used in subordinate clauses, compound sentences, and simple sentences. Notice the use of time markers such as "When I started medical school," "One day," etc.

2. Trace how the "plot" and the essay not only give information and perspective in a clear manner, but also add tension to the essay, keeping the reader's attention to the end. The combination of narrative and commentary work together very well.

Collaborative Activity

Ask students to prepare a mock trial of a student grievance hearing. Divide the class into the roles of defendants, prosecutors, judicial panel, and jury. The defendants are the students and professors who participated in the lab experiment; the prosecutors are students and professors who opposed the lab; the judicial panel asks questions of both sides; the jury listens and decides whether the dog lab will continue to be held and whether the students and professors are guilty of any wrongdoing.

Writer's Options

Assume the persona of a reporter asked to cover the dog lab. Write an article for the student newspaper of this medical school.

Multimedia Resources

Ask students to view the movie *Project X* with Matthew Broderick. This movie shows how military experiments on the effects of radiation from a nuclear blast used chimpanzees to test results.

1. The first school of thought accepted the "dog lab" as an opportunity to learn by observing the theory in reality. The second school of thought rejected the lab as unnecessary because they thought other ways of learning were available. The students do seem to have a choice. Whether they do or not depends on whether the instructor is credible when he tells the students they have a choice, or if the students think that their grades will be lowered.

2. McCarthy chooses to help anesthetize the dogs so she can be a full participant in the lab. She feels that her participation will guarantee the humane treatment of the dogs. Since her decisions have been unsure, it makes sense that her active rather than passive position would feel like the right course of action.

3. Once the experiment is almost over, McCarthy realizes that the cost of this experiment was not worth it for her. The experience gave her an opportunity to reflect on the person she was becoming, and the person she wanted to be. Because she is choosing to be a whole human, and to not just jump blindly through academic hoops toward medical school, she will be a better doctor. However, answers may vary.

Additional Questions for Responding to Reading

1. Do you agree that doctors are "different from other people"?

2. Are there other professions that can require a split between a person's personal values and professional values?

3. Are you willing to pay more for products that are not animal tested? Why or why not?

4. Would you be willing to wait longer for the approval of medical procedures or drugs so that animal life could be spared?

"DIRTY NEEDLE," NICHOLAS JENKINS

For Openers

Ask students to consider their own positions on the death penalty. Do they feel that it is not used often enough? That it is always wrong? Or are they somewhere in the middle, believing that sometimes it is appropriate? Ask them to place themselves on a spectrum within this debate. What types of arguments have they found most persuasive in reaching their conclusion?

Teaching Strategies

Carefully discuss the first paragraph with your students, making sure they understand the meaning and significance of each clause. What are the debates to which Jenkins refers? Do your students think he should have supported them more fully, or does his treatment seem appropriate for his audience?

Collaborative Activity

Ask students, working in groups, to find the key assertions that Jenkins makes in each section of this essay. Does he back up his points with evidence? Which points are substantiated, and which are not? As a class, see if there is a pattern. Are there places in the essay where your students want more information? If so, does this weaken his argument?

Writer's Options

Write a poem in response to Robert Lowell's. How do you see the aging gangster, waiting for death?

Multimedia Resources

To reinforce the historical grounding Jenkins provides, bring in some pen-and-ink drawings of executions from other eras. Sometimes they can be found in history or abnormal psychology textbooks.

Suggested Answers for Responding to Reading Questions

1. Jenkins deplores the "dream solution" because it allows people to keep an image of the death penalty that is quite far from reality. It absolves them of personal responsibility, and from finding other solutions within society for preventing crimes that merit execution.

2. Answers will vary.

3. Answers will vary.

1. Did this essay's title lead you to believe it would be about something else? If so, what? Given its possible ambiguity, is the title appropriate and effective?

2. Jenkins devotes most of the essay to providing historical anecdotes and other relevant information. How does understanding the history of the death penalty help readers accept his argument? What other strategies could he have used?

3. What argument does Jenkins put forth to justify why the death penalty should be abolished? Is that his main point?

"THE PERILS OF OBEDIENCE," STANLEY MILGRAM

For Openers

How does obedience function in our society? Does it preserve the social fabric, or does it actually corrupt or even undermine the ideals of a society? What examples can your students cite that might illustrate either of these views?

Teaching Strategies

Discuss the following with your students:

1. Let students know that Milgram was both praised and criticized for the study he describes here. Some critics called the experiment immoral and unethical because, they argued, the study subjected people to a degree of stress that could have had harmful psychological effects. Nevertheless, few could argue with the impact of the test results.

2. Remind students of the connection between Milgram's experiment and the killing of Jews and others in Nazi Germany, where soldiers committed unspeakable crimes and defended themselves by saying they were "obeying orders." Ask students at what point an individual has a moral obligation to say no, even if it means defying an order. You might also want to discuss the My Lai incident during the Vietnam War and consider what other options Lieutenant William Calley and the other soldiers had.

3. Milgram's essay seems to follow the general model of a scientific study. You might list for your students the various divisions of the formal science paper: introduction, identification of the problem, statement of the hypothesis, methodology, results, discussion, and conclusion. Would it have been as effective in another format?

Collaborative Activity

Ask students to work in groups to identify individuals whom they consider to be authority figures. Why are these individuals so perceived? Next, students should explore the circumstances under which they would be willing to defy those specific authority figures. As a class, discuss the different groups' conceptions of authority, and describe the conditions of possible resistance.

Writer's Options

Is "obeying orders" a valid excuse for carrying out an immoral or unethical act? Answer this question for yourself using your own experiences as departure points.

Multimedia Resource

There is film footage of Milgram's experiment, produced during the actual experiment; your school's psychology department or library may have a copy. Show your students the film footage, and ask for their reactions. Is it worse to read it or to see it? Why? You might also show the film *Judgment at Nuremberg*, which shows Nazi war criminals giving the "just following orders" defense.

Suggested Answers for Responding to Reading Questions

1. The dilemma inherent in submission is that when we defy authority we risk disrupting the social order. But submission allows us to abnegate our responsibility, which in turn allows corruption to be perpetuated. The experiments showed that ordinary people rarely have the inner resources to be defiant; most of us are anxious to please authority figures, and our desires to be obedient outweigh our sympathies with the sufferings of others.

2. Answers will vary. Some will say that we have been socialized to accept authority, and the subjects were able to deny their own

responsibility in inflicting pain. Other students will feel that the subjects of the experiment were weak individuals, and that they would have stopped the pain.

3. Answers will vary.

Additional Questions for Responding to Reading

1. According to Milgram, what is the dichotomy between conservative philosophers and humanists? In which camp would you place Milgram? In which camp would you place yourself? Explain your answers.

2. Why according to Milgram, would obedient subjects who believed they were inflicting pain consider themselves "on the side of the angels?"

3. What is the most extreme consequence of blind obedience, in Milgram's view? What pernicious effects does submission have on the individual who is submissive?

"THE RULES OF THE GAME," CARL SAGAN

For Openers

1. Ask students to recall a childhood experience that taught them some lesson, large or small, about being part of a society and taught them that personal sacrifice may be necessary to be part of a society.

2. Ask students to consider what laws or rules have changed in their own life time.

Teaching Strategy

The author uses several phrases that are worthy of further exploration. Brainstorm with the class on definitions or descriptions of "moral code," "code," "ethical behavior," "pragmatic," and "rules."

Collaborative Activity

Divide the class into groups. Each group will found a new country. This exercise will involve naming the country, sketching out a constitution in which three or four principles and laws are described, and creating a flag. After the establishment of each country, each group will try to recruit citizens through a presentation of their codes.

Writer's Options

1. As a continuation of the collaborative activity, each group will write one paper about their country. This paper is actually a constitution that sets forth the laws of the land and explains the rationale behind them. Each group will also write a description of their country's flag including an explanation of its symbols. Finally, the group will create an advertisement for their country in order to attract citizens. Be sure to identify the publication and type of audience for which the advertisement is intended.

2. Individually, students can write about why they would or would not like to leave their group's country to join another country. Predict what the new country will be like 300 years after its institution. How will laws be adapted? What changes will be necessary? What advantages and disadvantages are foreseen?

Multimedia Resources

1. View an episode of a science fiction program such as Star Trek, Voyager. What are the ethics of these star ship communities? What is valued? Discuss whether or not they are really an advancement of present day culture.

2. Explore and respond to the collection of Sagan quotations found at this site:
   http://www.oberlin.edu/~zrudisin/Discover/quotessagan.html

Suggested Answers for Responding to Reading Questions

1. This experience showed him that he must see beyond his own wants and consider the needs of others. He may have also learned that just because people possess something, they should not necessarily keep it exclusively for themselves. Moral codes, then, oblige us to treat others in the human community in particular ways. His story describes how these codes "regulate human behavior," and how his father was regulating his behavior to the needs of another.

2. Answers will vary. Some students will have religious rules that define moral behavior, while others will seek answers more on a

case-by-case basis. Sagan suggests humans share a basic goodness. However, cultures, conditions, and objectives shift. Therefore, Sagan does not suggest that there are any definite answers. The essay proceeds to illustrate how different rules emphasize and justify different behaviors.

3. The Prisoner's Dilemma demonstrates the need for trust and the consequences of breaking trust. Either trust or defection can lead to distress. "Real life" is more complex because it involves cultures, histories, and individual goals. Sagan suggests that the Prisoner's Dilemma teaches us the need for critical thinking. We must be able to make connections and choices based on knowledge of the multiple facets of human communities.

Additional Questions for Responding to Reading

1. Consider an ethical decision you recently made. By what rule would you categorize your choice? Is it disturbing to conclude that not all one's decisions follow the Golden Rule?

2. If the Golden Rule were always followed, what would happen to talk shows and newspaper reporting?

3. If "rules" are a result of inheritance and culture, consider the rules or codes that are passed on to you from your family. Describe your own family codes and rules.

Focus: Are All Ideas Created Equal?

Before reading either of the selections that follow, ask students to list the choices they make on a daily basis, even little ones. They should think first about those choices where they have two (or more) distinct options, each leading to a different set of consequences. Next, they should consider the choices they make that require little to no thought. What has become routine? What has become unquestioned in their lives? Do all actions involve choice?

"EQUAL TIME FOR NONSENSE," LAWRENCE KRAUSS

For Openers

Ask students to consider different debate styles and purposes in politics, the judicial system, and in academia.

Teaching Strategies

1. As a scientist, Krauss insists on logic and evidence as essential to any inquiry. Point out how all the assertions are supported by a specific example or reference.

2. Ask students to discuss the difference between evidence in a chemistry class versus evidence in a sociology or English class. How are "right" and "wrong" answers determined in the different disciplines?

Collaborative Activity

Ask students to make observational notes in their different classes on what constitutes evidence for the different theories presented. Ask students to also note some of the uncertainties that are created and explored in different disciplines.

Writer's Options

Carl Sagan's wife, Ann Druyan, says of her husband, "Carl never wanted to believe. He wanted to know." How would Krauss respond to Sagan? Using specific examples, explain the difference between "believing" and "knowing." Consider how each are effective and limiting in helping one come to a conclusion.

Multimedia Resources

Enter "creationism" as a key word search on your favorite search engine. Who is debating this issue? Examine the language used in the various articles. Determine the quality of the debate. Are most arguments making primarily emotional appeals? Notice what groups are not part of the debate.

Suggested Answers for Responding to Reading Questions

1. Answers will vary. Students may place ideas of spontaneous generation, cold fusion, and human flight in the "simply wrong" category.

2. One can argue that an "anything goes" attitude prevails. An ethic of tolerance to all ideas makes all ideas open to debate. Libstadt would agree that the debates exist. However, she would strongly disagree that such a debate is valid.

3. Racial supremacy, assisted suicide, and the fortifying of tobacco by the cigarette industry might be considered issues that are "subjects for debate" that have become tolerated.

Additional Questions for Responding to Reading

1. What current affairs could Krauss include in his essay if he were writing it today?

2. To what extent does one need "certainty" in understanding some ideas in order to debate the "uncertainty" of other ideas?

"DENYING THE HOLOCAUST," DEBORAH LIPSTADT

For Openers

In the first paragraph Lipstadt says the deniers' "strategy was profoundly simple." Ask your class to figure out what that strategy was. Next, ask them to think of ways that the deniers' strategies might be used against them, to promote Lipstadt's point of view.

Teaching Strategies

Discuss the First Amendment with your students. Read the exact wording to the class, and analyze what appears to be covered by it. Do your students agree with Lipstadt's assessment that the First Amendment is a red herring in this argument?

Collaborative Activity

In groups, ask students to choose a publication (for example, a specific school or community newspaper), and compose a policy outlining what kinds of advertising that publication will and will not accept. Remind them to make sure the policy reflects Lipstadt's concerns. Each group can present and explain its policy to the entire class. Whose policy is most flexible? The most inclusive? The most restrictive? The most permissive? What are the justifications for the different policies?

Writer's Options

Imagine you were on the newspaper staff when the request for the ad came in. What would you have done? Would you have argued for one position or another? Would you have left the decision to others?

Multimedia Resources

Bring in the campus newspaper for your institution, and if possible, copies of other college papers. Look through advertisements with your students. Does there seem to be a coherent policy about what is included and what is left out? What do most of the ads have in common? What seems to be missing? Do you think the paper you are looking at would be likely to run Smith's ad?

Suggested Answers for Responding to Reading Questions

1. Answers will vary

2. Answers will vary. Hilberg's criticism of Lipstadt does seem valid. After all, by publishing a book entitled Denying the Holocaust she appears to be perpetuating the idea that there is a real debate about it. On the other hand, she appears to have been disappointed by the refusal of one student newspaper to print a historian's refutation, so she clearly believes that ignoring the problem of Smith won't make him go away. Krauss hopes that readers will become more discerning in examining the credibility of media articles. He would welcome Lipstadt's book because it exposes the opposing view.

3. Answers will vary.

Additional Questions for Responding to Reading

1. Lipstadt repeats the word familiar throughout the first dozen paragraphs of the essay. Why does she do this? What is familiar about the ad and its language?

2. The woman, Haney, at Ohio State said she felt she had been "had" by Smith. Do you think she was? Explain your answer.

3. The argument in this essay builds around the reactions of various college newspaper staffs. Lipstadt is very careful to present multiple examples throughout this essay. Does she include too many? Do all of the examples help support her argument, or do they seem unnecessary after a while?